BIRDS, MEN and BOOKS

By the same author

East Anglia And Its Birds
A Century of Bird Books
Swallows

Birds, Men and Books

A Literary History of Ornithology

Peter Tate

Henry Sotheran Ltd
2, 3, 4 & 5, Sackville Street London W1X 2DP.

Birds, Men and Books

First Published 1986

ISBN 0 9508219 1 8

Printed in Leicester by Remploy Ltd.

Birds, Men and Books

CONTENTS

ACKNOWLEDGEMENTS

During the preparation and writing of this book I have received help and encouragement from many people: in particular, Dr. Ian Galbraith of the Sub-Department of Ornithology of the British Museum (Nat. Hist.) at Tring for kindly allowing me to use the Museum facilities, and Anne Vale, the librarian who, once again has assisted me with a great deal of advice and information. Susan Glenn, Assistant Archivist at the Smithsonian Institution for providing, so promptly, photographs of American Ornithologists, and Stephen M. Russell, Secretary of the American Ornithologists Union for much valuable advice.

I would also like to extend my thanks to Alan Parker who did much of the photographic work, taking, as always, endless trouble to obtain the best possible results. To Eric and David Hosking for providing the photograph of Roger Tory Peterson, and John Armitage who permitted me to use the photograph of Harry Witherby and F.C.R. Jourdain. I am especially grateful to Elizabeth Lack for allowing me to use a family portrait of her late husband, David Lack.

Finally I wish to thank Robert Kirkman and the staff of Henry Sotheran Ltd., and in particular Graham Nowland whose editing and advice has been of inestimable value throughout.

ILLUSTRATIONS

Plate 10. Henry Seebohm, Victorian Businessman, who travelled widely throughout Europe and Asia collecting birds and eggs. Henry Sotheran Ltd.

Plate 11. Harry Witherby, whose work as author and publisher was a major factor in the development of ornithology in Britain in the first half of this century. Shown here with F.C.R. Jourdain (left). John Armitage.

Plates 12 and 13. Witherby's Handbook established the concept of a detailed handbook and this is still a valuable tool for ornithologists. It was preceded by the rather rare Practical Handbook of 1920 which is shown top centre of plate 13. More compact successors are the two Popular Handbooks by P.A.D. Hollom. When The Handbook Of The Birds Of Europe is completed in seven volumes it will be the best single information source for birds of The Western Palearctic, a fit successor to Witherby's concept. Henry Sotheran Ltd.

Plate 14. Lord Walter Rothschild, an exceptional naturalist who built up a great museum entirely from his own financial resources. BBC Hulton Picture Library.

Plate 15. John Gould the originator and publisher of a whole series of magnificent bird books. Henry Sotheran Ltd.

Plate 16. Richard Kearton. A pioneer bird photographer, who also wrote and illustrated many books. BBC Hulton Picture Library.

Plate 17. Roger Tory Peterson's name is synonymous with the modern fieldguide. Eric Hosking. FRPS.

Plate 18. David Lack, particularly notable for his outstanding work in the fields of bird behaviour and population studies. Elizabeth Lack.

Plate 19. The B.T.O. Breeding Bird Atlas which marked the advent of the era of detailed distributional studies. With permission from British Trust For Ornithology.

Plate 20. The concept of the fieldguide, originally published by Rodger Tory Peterson 1934, has today expanded to cover almost the entire world, the range shown above being a very small selection. Henry Sotheran Ltd.

I

The Primitives

That birds have been of interest to men since time immemorial is proved by the existence of recognisable images of them in prehistoric cave drawings, Egyptian frescoes and Babylonian bas-reliefs. Yet the occurrence of any serious literature upon the subject is, when measured beside this timescale, comparatively recent. Indeed as far as Britain is concerned it was not until the end of the 17th century that what could accurately be described as a true bird book first appeared.

It is not hard to account for this. The absence of books on the subject made it impossible in many cases to properly identify birds that were observed; then there were so many local names for common birds that, not surprisingly, the utmost confusion reigned.

Accurate field identification is no easy matter, when birds, particularly the smaller species, are timid or restless. Before the invention of binoculars in 1859 the only way to get a good look at a bird was to trap or shoot it, and reliable firearms were not developed until comparatively recent times. It was the sheer difficulty then in obtaining the most basic knowledge that inhibited the advance of ornithology as a science. This can be especially appreciated by comparing the subject to other branches of Natural History; botany for example, where the subject needs only to be approached, or entomology, where the species have limited ability to evade close study, though perhaps entomologists might disagree.

Apart from the practical difficulties attending the study of birds and nature, in medieval times an even stronger deterrent lay in the mental attitude engendered by an all-powerful church towards God's creatures. Without going too deeply into medieval religious philosophy some insight into contemporary attitudes to the natural world is necessary to a history of ornithological literature. The church's attitude is best summarised by the late Canon Charles Raven as follows: "But for the Catholicism of the West other-worldliness consists only in insisting that this world derives from and depends upon another; that it is the symbol and instrument of that other; and that it is to be studied and interpreted, if at all, not for its own sake but in order to disclose its supernatural meaning". This belief confined thought and investigation to an extent that it is difficult for us to appreciate today.

The church saw life as a pyramid at the apex of which stood man, and above this, God the Creator with the saints and angels. From its early days the church had always used an allegorical interpretation of the life of animals, treating them as fables for the instruction of man. The cunning of the fox, the industry of the ant and the patience of the ass were held up as an example to be either avoided or copied. In this sort of intellectual climate to contradict the church's view of the true meaning of the animal world, or to question the biblical evidence that God created each species separately and immutably, was not only wrong but heretical. The legacy of this line of thought was, as late as 1860, to lead the church to oppose the theory of evolution as propounded by Darwin and Wallace. Undoubtedly these beliefs were deep and sincere and although in the light of later knowledge they appear to be wrong the lives and works of our predecessors should in fairness be judged within the context of their own times and not of ours.

A revealing illustration of the attitude of the church to animal life in medieval times can be quoted from Giraldus Cambrensis: "in the birds of prey, the female is always the larger, and that signifies that in every sort of evil doing the females are stronger than the males". Although better known by his latinized name, Giraldus Cambrensis was born Gerald de Barri at Manorbier in Pembroke sometime around 1146. He was educated in Paris and ordained a priest in 1172. Appointed to collect tithe monies from the Welsh he eventually became Archdeacon of Brecknock. In 1184 he became chaplain to Henry II and accompanied his son, Prince John, on an ill-fated expedition to Ireland. This journey was to result in "The Topography of Ireland: Its Miracles and Wonders", a manuscript written in 1207 and eventually published in 1587 together with Holinshed's Chronicles, but sometimes found separately. Two chapters of this work are concerned with his observations on birds and their habits. He was an observant and intelligent man who recorded many useful pieces of information about the natural world about him but capable of including the wildest of fables in his writings. For instance he could clearly separate the mute swan from the whooper, but then proceeded to state categorically that he had seen barnacle geese hatching from barnacle shells on tree trunks in the sea. He also firmly asserted that storks spent the winter at the bottom of rivers. He was a conceited and pompous man much given to spiteful quarrels, hence much of his ecclesiastical life was spent in strife and petty religious arguments. He was proposed as bishop to the See of Lincoln, but his cantankerous nature had brought him many enemies and his election was strongly opposed. It would seem however that he spent the closing years of his life in some peace and retirement and died sometime around 1216.

Although he could hardly be properly described as a naturalist Giraldus' observations on birds exemplify the kind of scattered ornithological material which appeared from time to time in church records and manuscripts. Throughout the medieval period various haphazard references to birds occur in documents and treatises concerned with matters of general interest as well as comments on birds contained in church records and religious writings.

A more likeable character, and almost an exact contemporary of Giraldus was Alexander Neckham. Born in 1157 at St. Albans he was educated in that city proceeding thence to Paris where he studied at the school of Adam de Petit

Pons and by 1180 was a distinguished lecturer at the University. In 1186 he returned to England and was placed in charge of the school at Dunstable. He is notable for being the author of one of the earliest works on natural history to be written in England. Entitled "De Naturis Rerum" this is a compendium of religion, astronomy and natural history and mythology, many of the creatures he describes such as the grips, phoenix, basilisk and mermaid being quite imaginary. He probably copied the overwhelming majority of the book from the works of ancient writers such as Pliny the elder, Carius Julius Solinus, Virgil, and a rather more recent author of the 6th century called Magnus Aurelius Cassiodorus. This practice of borrowing and copying from other mens' work was by no means confined to Neckham. The custom was widespread both here and on the continent generally throughout the medieval period. Like Giraldus Alexander Neckham could not be described as an ornithologist in any serious sense but he is worthy of consideration. For his time his range of knowledge was extraordinarily wide. Known also by his latin nickname 'Nequam' (wicked) he became Abbot of the Augustinian abbey at Cirencester in 1213. He died in 1251 and was buried in Worcester Cathedral. "De Naturis Rerum" remained in manuscript form until it was eventually published in the Rolls series of chronicles and memorials in 1863.

Brief and fragmentary as they are these early ornithological references are of no great value but little pieces of information can be interesting. For example, Matthew Paris, a monk at St. Albans Abbey, writing in 1251 gives an excellent account of an invasion by crossbills, going into some detail as to how they damaged the fruit in the orchards by eating the pips of the fruit and thereby ruining the crop. This is an interesting and curious observation in that as a rule crossbillls only rarely attack fruit, their primary food source being the seeds from fir cones.

One of the earliest printed books on natural history in the English language, (although the original edition had been printed in Latin in Cologne by Caxton in 1472), was entitled "De Proprietatibus Rerum" by Bartholomaeus Anglicus. This work was translated from its original Latin into English by John of Trevisa for Sir Thomas Berkeley in 1397, and was first printed in this form in London around 1495 by Caxton's apprentice Wynken de Worde who took over the press. Barthlomaeus Anglicus was born in England around 1230 and became a minorite friar and a professor of theology in the University of Paris. His book was divided into nineteen parts one of which was devoted to birds. Thirty-seven species were listed, some of which were rather vague and uncertain. He is believed to have died around 1260, so his work (like that of Giraldus Cambrensis) was not published until about two hundred years after his death.

A whole field of ornithological literature of considerable importance in the middle ages was that concerning falconry. Besides its primary subject matter, there were frequent references to birds other than hawks. One such book which is amongst the first books in England is "Book of St. Albans, containing Treatises of Hawking, Hunting and Cote Armour" published in 1486. This is believed to have been written at least in part by Juliana Berners, at one time Prioress of Sopwell Nunnery in Hertfordshire. She is credited with having written, (a) a small portion of the treatise on hawking, (b) the whole treatise on hunting, (c) a short list of the beasts of the chase and, (d) another short list of beasts and fowls. Although rather a borderline bird book which hardly advanced ornithological knowledge, it does, owing to the rarity of any books on this subject at this time, merit a place in our history. Only one copy is known of the original, but a facsimile was published in 1881.

The arrival of the Tudor dynasty and a new age coincided almost exactly with the advent of the first ornithological writer of importance. William Turner was born at Morpeth sometime around the year 1500. The son of a tanner, little is known about Turner's early life before he went up to Pembroke Hall in Cambridge in 1526. Whilst at Cambridge he became a close friend of Nicholas Ridley and Hugh Latimer, both of whom were to be burnt at the stake for heresy. Turner's own strong

Protestant beliefs were later to force him to seek refuge on the continent from religious persecution. Graduating in 1529 he became a fellow of Pembroke Hall in 1531 and its senior treasurer in 1538. His deep interest in birds developed during his years at Cambridge, and in particular for those to be found in the fens. At that time the fens were largely undrained, the fenland was far more extensive than today and reached considerably nearer to the actual town of Cambridge. With its vast areas of quaking bogs interspersed with small islands of drier ground often enveloped in thick miasma and mist it must have been a forbidding place yet, like so many who have come to know that unique area, Turner seems to have been fascinated by its very strangeness. Certainly in his time it contained huge numbers of birds including many species, such as cranes and black terns, which no longer breed there.

Around 1540 William Turner left Cambridge and led an itinerant life travelling constantly across England preaching. Whilst in Oxford he was imprisoned for preaching without a licence. Probably his detention was the result of his radical religious beliefs rather than the lack of a licence. Upon his release he and his wife left England for the continent. Whether he was officially banished is not known neither do we know precisely where he went. Certainly he visited Bonn, Cologne and Basle and eventually reached Italy where he stayed in Bologna studying botany under Luca Ghini. He took his doctorate of medicine either in Bologna or possibly Ferrara. Travelling to Switzerland in 1543 he met Conrad Gesner in Zurich. Gesner was one of the greatest naturalists of the 16th century and he and Turner soon became close friends and continued to correspond a great deal over the years. By 1544 Turner had returned to Cologne where he published his major work on birds, "Avium Praecipuarum Historia". This he dedicated to the young Prince of Wales, later to become King Edward VI. This work is important not only because it is one of the earliest bird books to be published, but because in it the author made one of the first serious attempts to separate mythology and folklore from reality. In addition he made every effort to introduce accuracy and order into the naming of birds and flowers. Whilst he did continue to repeat some of the old fallacies, many of his own observations were original and accurate. In all he described some 130 species of birds of which 25 were domesticated species or species foreign to England, but at least fifteen species were mentioned for the first time in a written work. A reprint of this work was issued by Mr. George Thackeray in 1823, and in 1903 A.H. Evans of Cambridge published a further reprint with a translation.

In 1544 Turner undertook the editorship of "Dialogue de Avibus" which had been written by Gybertus Longolius, who lived in Utrecht but had died before its completion. On the death of King Henry VIII Turner was able to return to England, the Protestant religion now being firmly favoured by Edward VI. He became physician to Edward Seymour, Duke of Somerset (the Lord Protector) and in 1550 was appointed Dean of Wells. Unfortunately for him Edward's reign was brief and on the accession of the fiercely Catholic Mary Tudor, he was forced to leave England again. Again precise details of his travels are lacking but it is probable he returned to Cologne and then to Worms. Much of the evidence for his travels come from references to plants found in various places and recorded in his "Herbal".

On the death of Queen Mary in 1558 he was free to return to England but experienced some difficulty in regaining his position at Wells. But in 1561 he finally obtained a writ to force the incumbent, Goodman, to vacate the deanery and allow him to return. Whilst awaiting his reinstatement to Wells Turner continued to do a lot of botanising and made numerous revisions to his "Herbal".

He would seem to have been a blunt, uncompromising man and this was to lead to a great deal of friction between himself and his bishop over such matters as the wearing of the correct ecclesiastical robes. It would be true to say that as he grew older William Turner regarded himself more as a doctor than a cleric. Evidence of this is given by his habit of adding after his signature, 'a physician delighting in the study of sacred literature'. He died at his house in Crutched Friars in London on 7th July 1568. Although he had expressed his intention of writing more on birds and fishes in the future the main part of his sparetime during the latter part of his life was devoted to his "Herbal" and other botanical papers. Undoubtedly he possessed a fine and scientific mind and he can be considered as one of the true pioneers of natural history in England.

Amongst William Turner's contemporaries at Cambridge was John Kay (Caius). Born in Norwich in 1510 he went up to what was then called Gonville Hall in Cambridge in 1529, where he read classics and religious studies but was not ordained. After spending several years translating various theological works from their original Latin and Greek he went to Padua in 1539 in order to study medicine under Andreas Versalius, the great anatomist of the day. Kay qualified as a doctor of medicine in 1541 and returned to England in 1544. Elected as a fellow of the College of Physicians in London in 1547 he was elected to the office of President no less than nine times. He was also appointed as physician both to Edward VI and Queen Mary. John Kay was always keenly interested in natural history and this was to be expressed in his publication in 1570 of "De Rariorum Animalium atque Stirpium Historia". Unfortunately, regarding Natural History, he lacked a truly enquiring mind and was prone to repeat fables and stories as told to him without too much critical examination. He also had a marked tendency to get sidetracked into rather abstruse theories about the derivation of words to the detriment of the study of the object so named. He was first and foremost a doctor and only secondly a naturalist and the amount of original work he produced was therefore limited. But at a time when naturalists were few we should not underrate his contribution.

Kay certainly knew William Turner but all the available evidence suggests that the relationship was cool. Kay was a devout Roman Catholic which at this time of intense religious feelings would have immediately estranged him from the reformer Turner. He was a bachelor, highly studious, rather humourless and apt to adopt a high-handed attitude to other people. Not a particularly likeable person he felt himself to be unappreciated and misunderstood; he died in July 1573 an embittered man. He left all his wealth to his college which, as a result, designated him its co-founder, the college being renamed Gonville and Caius.

A rather lesser character, but still worthy of consideration was Edward Topsell who was born in Sevenoaks in 1572. He went up to Christ's College, Cambridge, where he was ordained and then held a number of livings mainly in Sussex before becoming the curate at St. Botolphs, Aldersgate in London. He is best known for his "Historie of Four-Footed Beastes" published in 1607, and the "Historie of Serpents" which appeared in 1608. Both these books consisted mainly of extracts or, to be more accurate, copies of the works of Gesner which he translated into English. His few original observations were somewhat trivial and not too reliable. He is primarily of interest because he wrote one of the first books on birds to be written in English, namely, "The Fowles of Heaven, or Historie of Birds". This again was mostly a rehash of someone else's work, in this case, that of Ulysse Aldrovandi. Topsell's book was never completed and had not been published when he died in 1625. The manuscript eventually came to rest in the Henry E. Huntington Library in the U.S.A. and was finally published in 1972, three hundred and fifty years after it had been written. Some 124 birds whose names begin with the letter A, B and C were covered each being illustrated by a coloured drawing which are interesting examples of the illustrative art of the time. Painted in water colours and outlined in Indian ink the drawings range in size from two by three inches to five inches square and each one is pasted onto the page close to each bird's respective description. Several are quite unidentifiable and some highly fanciful, such as the birds of paradise which were drawn without legs simply because it was believed that they stayed in the air for the whole of their lives. This particular myth had arisen because the birdskins brought back to Europe had had the legs removed during the preservation process, so people simply assumed that they had never possessed such limbs.

Whereas the work of Edward Topsell represented, if anything, something of a regression in ornithological knowledge, that of Sir Thomas Browne presented a total contrast. Here was a true pioneer of the development of ornithology in this country, whose work added greatly to the knowledge of wildlife in general and birds in particular. He was one of the first naturalists who could truthfully be described as a genuine field observer. Unlike all his predecessors he did not rely on quoting the observations or imaginings of bygone writers but carefully watched and studied his subject for himself. Born in London in 1605 Thomas Browne went up to Pembroke College (then called Broadgate Hall) and then went to Leyden where he obtained his doctorate in medicine in 1633, returning to live in Norwich in 1636. He was a man who was interested in every living thing that he saw around him, noting the habits of fish and invertebrates as well as birds. A generous and ingenuous man he seemed quite uninterested in gaining credit for this new found knowledge and happily shared it with others without any thought of acknowledgement. Many of his observations have come down to us by way of his letters to other naturalists, in particular Dr. Christopher Merrett and the great John Ray, both of whom frequently noted that they had received important help from "our Norwich correspondent". In addition to many medical works Thomas Browne wrote "Pseudodoxia Epidemica" which was first published in 1646. This is a great treatise on life in general and his thoughts on it. This book clearly reveals him as a man who had wholly freed himself from the old medieval beliefs and superstitions and as the possessor of an enquiring and agile mind. He was the first naturalist in Britain whose outlook could be described as "modern". At a time when belief in hibernation was almost universal Browne firmly stated that birds migrated to warmer climes. By means of numerous dissections he greatly enlarged the available knowledge of the anatomy and physiology of birds and mammals. In addition to his primary interest in zoology he also possessed a considerable amount of botanical knowledge, a wide knowledge of plants and their curative properties being an essential part of any medical training. Many of his observations were published in 1902 as "Notes and Letters on the Natural History of Norfolk", edited by Thomas Southwell. The great importance of Sir Thomas Browne as a naturalist lies not so much in his publications, for many of his discoveries were published by others, but for his progressive attitude of mind which was a pointer to the future of ornithological thought. He died in Norwich on his birthday in 1682.

His close friend and correspondent, Christopher Merrett, was born at Winchcombe, Gloucester in 1614. Like Sir Thomas he was a doctor of medicine, going up to Oriel College in Oxford in 1631. He was a popular figure and pursued a highly successful medical career becoming a fellow of the Royal College of Physicians in 1651. In addition he was appointed curator of the Harvey Museum and Library. In 1662 he became a founder member of the Royal Society. His book "Pinax Rerum Naturalium Britannicarum" published in 1666 listed some 170 species of birds of which eight were described for the first time. His zoological work was more in the nature of a compilation, culled mainly from the works of Johnstone, Gesner and Aldrovandi, and contained little original material springing from his own observations. He started out with the admirable intention of excluding from his lists any item which he considered to be of doubtful origin or founded on hearsay. He failed signally to do so and a number of highly dubious accounts crept into his writings. He was rather more knowledgeable in matters relating to geology than zoology or botany but even in this subject lacked sufficient basic knowledge to evaluate properly the material which was submitted to him.

It is difficult for us who have always known a virtually stable naming system for birds, beasts and flowers to imagine the chaos into which nomenclature had sunk by this period, immediately prior to the work of Linneaus. Practically every authority had its own order and criteria on which its groupings were made, and a single species might be placed in different orders, with a different name, in lists which were published at more or less the same time. Even today there is some degree of switching species to different genera and some minor nomenclatural changes, but at least such alterations are made according to an agreed set of rules, so it is relatively easy to keep track of the changes. To illustrate the appalling mess which Merrett made of the classification of birds, he grouped them in the following highly arbitrary fashion: — "Land birds carniverous", includes hawks, the cuckoo, owls and crows, sea pie (oystercatcher), the bat, the shell apple (crossbill) and the goatsucker (nightjar). "Seed-eaters not songsters", gamebirds including snipe, woodcock and hoopoe. "Dust-lovers that wash" are coot, kingfisher, pigeons, house sparrows and reed sparrows. Finches and larks are classified as "Seedeaters — songsters", while thrushes, starling and clotbird (wheatear) are lumped together as "Berry-eaters"."Insect eaters" include woodpeckers, swallows, tits, chats, nightingale and dotterel. Under "Waterfowl — Web footed" are designated geese, capercaillie, ducks, and divers gulls and under the title "cloven-footed" are storks, herons, sand-pipers, plovers, and corncrakes etc.

By virtue of his training and position Merrett had the chance to advance significantly the course of natural history in England, but he proved to be quite unequal to the task. The opportunity to build upon the foundation laid by Sir Thomas Browne and to consolidate the trend towards new thoughts as set forth by him was lost. His last years were sad and desolate as one catastrophe followed another. He lost his library and collections and many copies of his "Pinax" in the Great Fire of London in 1666, a disaster which also destroyed the library and museum of the College of Physicians. Then Merrett proceeded to launch a series of attacks upon the apothecaries. He wrote a paper entitled "Short View of the Frauds Commited by

Apothecaries'' which immediately embroiled him in a bitter and heated controversy with another physician, Henry Stubbe. There followed an argument with the College of Physicians over his position in the Society and his stipend as its curator. The final result of all this unpleasantness was that he was expelled from the Society, and although he lived on until 1695 he did not do any more work of consequence. He was buried ''Twelve feet deep'' in the church of St. Andrews in Holborn.

At this early period there were no contributions to ornithological literature from elsewhere in the English speaking world since colonization was relatively recent in the Americas and the Australias had yet to be discoverd. The early settlers in Northern America, although they had little time or opportunity to write about them, did observe that turkeys, passenger pigeons and waterfowl were exceedingly plentiful and provided food. They also noted that after less than a century of colonization they were already becoming noticeably less abundant. Some species were commented upon, particularly those who were given the names of the European birds which they might superficially resemble such as robin, blackbird, sparrowhawk etc. The first account of birds from North America must be that of Christopher Columbus who in his journal for October 8th 1498 recorded that ''There were many small land birds and they took one which was flying to the south west. There were jays, ducks and a pelican''.

An important class of ornithological books which is of great interest to many people is that devoted entirely to the birds of a particular local area. These may be large county avifaunas or parish avifaunas covering much smaller districts. In the United States the same pattern has been followed and every state has its own bird book, as have many cities and national parks.

In Britain the first attempt at such a book was ''The Natural History of Oxfordshire — being an essay toward the natural history of England'' published by Robert Plot in 1677. It was a very incomplete book which did little towards the advancement of any knowledge concerning the birds of Oxfordshire. Robert Plot was born in Kent in 1641 and went up to Magdalen Hall,

Oxford, his chief claim to fame being that in 1683 he was appointed the first keeper of the Ashmolean Museum, and in the same year was appointed Professor of Chemistry to the University. In 1694 he was nominated Mowbray Herald and died in 1696.

The conception of a book covering a specific area in some detail was soon to be repeated though and in 1700 Charles Leigh published a book entitled "The Natural History of Lancashire, Cheshire and the Peak in Derbyshire, with an account of the British, Phoenician, Armenian, Greek and Roman Antiquities in those Parts". In view of the wide range of his subject it is hardly surprising that the treatment of the birds was somewhat trivial!

It can fairly be said that until the latter half of the 17th century no book had been published which could rightly be classified as a bird book, as we understand the term today. However, as far as Britain was concerned, the work of one man, John Ray, was to change the picture entirely.

II

First Scientific Approaches

John Ray was perhaps the greatest all round naturalist that Britain has produced, and certainly the greatest of his age. Throughout his career he worked to dispel the medieval heritage of ignorance, superstition and chaos in the natural sciences which had restricted all his predecessors, most of his contemporaries, and which was to afflict some of his successors. By the end of his life Ray had established the direction and general principles which marked the emergence of ornithology as a serious scientific discipline.

He was born in 1627 at Black Notley in Essex, a blacksmith's son. His mother was the village herbalist and he received from her an excellent grounding in botany. His intelligence was quickly recognised and the village priest, Joseph Plume, obtained a place for him at Braintree Grammar School, where he excelled particularly in Latin, which was to be of vital importance to him in later life. Thanks to the assistance of another clergyman, the Rev. Samuel Collins, Ray was admitted to Trinity College, Cambridge, transferred to Catherine Hall and then returned to Trinity, graduating from there in 1648. At Cambridge Ray read the classics, Latin, Greek and Hebrew, becoming fluent in the first two. He also had a fair command of French and Italian besides being a good mathematician. A true polymath and a brilliant scholar his mind was as remarkable for its total lack of cant or dogma as for sheer breadth of knowledge.

Much of his early work such as his journeys around Britain and Europe lie outside the scope of this book, they were devoted primarily to the collection of specimens for his botanical works, the "Catalogus Angliae" and "Historia

Plantarum''. Nevertheless, given his active and enquiring mind, the many hours spent in the field obtaining this information must certainly have added to his ornithological knowledge.

John Ray's greatest contribution to natural history was entitled, paradoxically, ''The Ornithology of Francis Willughby. Translated into English and enlarged with many additions etc. by John Ray'' published in London in 1678. Ray's acknowledgement in the title to his friend was to contribute to a tasteless controversy two centuries later.

Although this book bears the name of Francis Willughby by far the major part of it was the work of John Ray. He and Francis Willughby had collaborated for many years, travelling and working together on several expeditions collecting botanical specimens and observing wild life. Following the untimely death of Willughby in 1672, the preliminary notes for a book on birds which he had planned were given to Ray. Besides being collaborators the two men were close friends and on Willughby's death Ray not only received a legacy from him but also acted as his executor.

Ray had no independent means. In 1660 following the re-enactment of a University Ordnance, suspended during Cromwell's rule, that fellows must be ordained within seven years of receiving their fellowship, he was ordained a priest. In 1662 with Charles II firmly back on the throne the Act of Uniformity was passed which required all priests to acknowledge the absolute authority of the established church. The High Church tenets then favoured by the Establishment did not appeal at all to the puritanical Ray and he refused to swear the necessary oath. This placed him in an impossible position since he was consequently no longer accepted as a priest but since he had been ordained he was debarred from pursuing any secular occupation. It was now, at the age of 35, that he decided to devote himself to the study of natural history.

The notes which had been left to him on Willughby's death were mostly descriptions of birds' plumages, and formed only a meagre basis from which Ray was to develop the first genuine bird book to be published in Britain. Along with his wide knowledge of birds, Ray brought an enquiring and scientific mind to a subject previously bedevilled by a muddled and slapdash approach. Unlike most of his predecessors he had considerable field experience, and he had seen many of the birds which he was now to describe. For example, on his journey to Wales on a visit to Puffin island in the Menai Straits he noted that he had seen, "two sorts of seagulls, cormorants, puffins, so called there, razorbills, guilloms, scrays two sorts of which are a kind of gull", (terns). On a visit to Bardsey he recorded that he saw, "prestholm puffins, seapies and some other birds". The bird which he called a prestholm puffin, is the bird which we call simply a puffin today, whilst the puffin referred to in the Isle of Man is clearly the Manx shearwater.

In 1661 he visited the Bass Rock in the Firth of Forth and wrote how he saw "the scout, which is doubled ribbed (razorbill), the cattiwake, the scart (shag) and soland geese (gannet)": and reported that "they sit in multitudes till one comes close up to them". On another journey he noted that near Padstow "we saw great flocks of Cornish choughs". Visits to other countries provided him with further material for his book, and he recorded numerous species of birds which he found in markets in Italy, Sicily and Germany. He watched cormorants nesting in trees at Lekkekirke near Rotterdam, and night heron, spoonbill and grey heron nesting in the reedbeds on Texel — much as they still do today.

His interest in birds was not confined to observation, as his letters to Martin Lister, the zoologist, clearly reveal he also made careful studies of their anatomy. He discusses the specialised musculature of the woodpecker's tongue which enabled the bird to greatly extend its tongue in order to extract grubs from holes in rotten wood. He was much intrigued by the structure of the trachea in the shelduck, and was aware that this organ differed between the sexes. This prompted him to dissect a red-breasted merganser in order to see if the same formation of the trachea was also present in that species.

In the "Ornithology" any species about whose existence he had doubts were relegated to an appendix and, unlike most of his contemporaries, he omitted such fanciful birds as the phoenix, griffin and harpy. Professor Alfred Newton writing in 1893 was correct in his assessment that the "'Ornithology' was the foundation of scientific ornithology in Britain".

Following the "Ornithology" Ray wrote "The History of Fishes" in 1686, a book which had been planned by Willughby. In 1693 he followed this with "Synopsis Quadrupendum", and also published revised editions of his earlier works. He also wrote a number of philosophical works, "The Wisdom of God" in 1691 and "Persuasive to a Holy Life" in 1700.

Ray passed his final years in Black Notley having returned there in 1679 after the death of his mother. Here he was visited by many eminent men of science including Sir Hans Sloane. He died at his home 'Dewlands' on 17th January 1705. Sadly most of his papers were lost and much that we would like to know about this remarkable man disappeared along with them. As a final illustration of the breadth of Ray's outlook, he was amongst the first to express strong views on the harm done to biological study by the division between the theoretical laboratory orientated type of approach as opposed to practical observation in the field. The difficulty has not been completely resolved today.

Willughby's reputation was overshadowed by John Ray, although the former's relative obscurity is largely due to his early death at the age of 37. He was born in 1653 at Middleton in Warwickshire, and in 1653 went up to Trinity College where he met Ray, and later provided Ray with a number of botanical records which he incorporated in his "Catalogus Plantarum circa Cantabrigiam Nascontium". He did not join Ray on any of his early excursions but in 1660 accompanied him on a journey through the north of England and the Isle Of Man. Their precise itinerary in unknown but they did visit Sheffield and Settle in Yorkshire, from there they went to the Lake district around Kendal and also visited parts of Lancashire.

This trip firmly cemented the friendship between the two men and also marked a turning-point in Ray's career. From now onwards he became a zoologist rather than a botanist. Their longest expedition together was undertaken in 1663 when in the spring they set out from Dover for France accompanied by two other friends, Philip Skippon and Nathaniel Baron. The sudden oubreak of war in France forced them to ride on straight to Flanders away from the action. From there they proceeded south to Frankfurt and onwards to Switzerland, Italy, Sicily and Malta. On their return to Montpelier (in France) the party divided and Willughby went alone to Spain not returning to England until December 1665 on the death of his father. For a while family business prevented him from travelling far, but he and Ray did manage later to visit Worcester, Gloucester, Dorset and Cornwall on a botanical collecting expedition. He now began to contemplate seriously a journey to America, but at this point his health, never robust, began to fail and he died on 8th July 1672.

In the mid nineteenth century a controversy emerged to dispute the relative contributions to the "Ornithology" made by Willughby and Ray. Some implied that Ray took much of the credit for the work at the expense of his deceased friend but this would seem to do him a great injustice. Sir James Smith, an eminent 18th century botanist who was well qualified to judge, stated that John Ray was "The most accurate in observation, the most philosophical in contemplation, the most faithful in description of all botanists of his own or any other time". Canon Raven in his biography of Ray wrote that "his university record bears out this testimony — he was a man of outstanding ability amounting to genius. He was scrupulously punctilious, indeed generous to a fault in acknowledging the help of others. Witness his insistence in his botanical works on acknowledging the help of others, his insistence in his botanical works on stating that this plant was first discovered by Mr. M. Lister and shown to me, this was found by Mr. Johnson and shown to me, this was shown to me by Mr. T. Willisel — and so forth".

To quote Canon Raven again:- "There was considerable evidence that there was a whole-hearted partnership between the two men, and to decry one or the other is to do an injustice to the friendship". Again "If it is necessary to categorise them the evidence makes it certain that Ray was a scientist of genius, and probable that Willughby was a brilliantly talented amateur. There is no evidence at present that they were in the same class as naturalists". So far no evidence has appeared to refute that commanding conclusion.

Although John Ray laid the foundations of scientific ornithology many writers who followed were not so particular checking their sources, or engaged in much original research. Fiction continued to be recorded as fact. They were of course still restricted by a lack of basic knowledge from which to check new theories and provide guidelines on which to base new beliefs. For example, although a number of people strongly suspected that many birds migrated in order to escape the winter, there was no way by which this could be proved scientifically. Ringing birds with light metal rings was not yet practised and there was an almost total lack of reliable observers in any areas where birds might winter. Nearly two centuries were to pass before irrefutable proof of migration was forthcoming. Even Gilbert White believed that at least some birds hibernated during the coldest months, but was too perceptive to deny his suspicions that some species did leave these shores in order to avoid the winter.

The widening of interest in natural history which occurred during the 18th century manifested itself in the formation of societies dedicated to the study of its various branches. A major contribution to this blossoming of interest was the development of cheap printing. As access to the discoveries and learning of others became faster and easier, knowledge was no longer confined exclusively to a privileged minority. Today when information is freely available to an almost embarrassing degree, it is difficult to appreciate the intellectual wilderness in which the vast majority of the population formerly existed.

The first natural history society of which we possess definite knowledge was "The Temple Coffee House Botanic Club" founded around 1689. "The Society of Aurelians" which concerned itself with the study of insects was established sometime around 1740. But no society whose interests lay primarily with birds was formed until the second half of the 19th century, and in Britain "The Mammal Society" was not founded until 1954. The reason for the early pre-eminence of botany was simply the practical importance of herbs in medicine: the apothecary's social status was similar to that of today's general practitioner.

The first half of the 18th century saw almost a collapse of science in Britain. Sir Isaac Newton died in 1727, and thereafter the Royal Society deteriorated to an alarming extent. At Oxford in 1730 the president of Trinity, one Huddesford, was elected keeper of the Ashmolean at a salary of £50 per annum. "whether he do anything or not". Humphrey Sibthorp, Professor of Botany for thirty-six years, published not one scientific work, and as far as is known delivered only one lecture, and that an exceedingly poor one. Things were no better at Cambridge where a man called Martyn held the chair of botany but was absent for thirty years, only resigning it in 1792 in order that his son might succeed.

Corruption and graft were everywhere. Following the rise to power of the Duke of Newcastle and later Walpole, bribery and nepotism were rife, and positions in the army, navy and government were reserved for those prepared to support the ruling Whig party. In such a climate it was hardly surprising that the Universities stagnated along with the rest of society. Despite this general lack of direction and effort there were still some who continued to further the course of natural history and in particular, ornithology.

One such was George Edwards who was born at Stratford in Essex in 1694. After travelling widely in Norway, Holland and France he returned to London and commenced a lengthy correspondence with the great Swedish naturalist, Linnaeus. In 1731 he was recommended by Sir Hans Sloane for the appointment of librarian of the Royal College of Physicians and in 1750 he was awarded the gold medal of the Royal Society, later becoming a fellow of that august body. He is best remembered today for two major books on natural history. The first, "A Natural History of Birds" was published between 1743 and 1751. This was issued in four parts and contained 52 copper plates depicting sixty birds and two mammals. None of the drawings was particularly accurate. His second and greater work, in effect a continuation of his first book was entitled "Gleanings of Natural History, exhibiting figures of Quadrupeds, Birds, Insects, Plants etc. with descriptions" and was published between 1758-64. This contained three hundred and sixty-two coloured plates. The book went through numerous editions and collation is difficult since several printings were often combined in one copy.

Unlike Ray, George Edwards was not a scientist, but he was an honest, accurate and conscientious observer of birds and mammals. Swainson described him as having "the simplicity and piety of Isaac Walton, and may be looked upon as one of our greatest worthies". His books were considered to be of sufficient merit for them to be translated into French, German and Dutch. Around 1764 he went to live in Plaistow, Essex and died there, of cancer, in 1773 and lies buried in West Ham. Through Edwards there is a connecting link between the early literature of Britain and North America, for he edited an edition of Mark Catesby's "The Natural History of Carolina, Florida and the Bahamas: Containing the Figures of Birds, Beasts, Fishes, Serpents, Insects and Plants etc." published in 1754.

Mark Catesby was born around 1679, probably in London. He studied natural science and in 1710 managed to find the means to visit the New World; where he travelled widely and in 1719 returned with a fine collection of specimens. This particularly delighted Sir Hans Sloane as they were by far the most comprehensive collection of skins to have been so far brought to Britain from America. Catesby returned to America in 1722 and settled for a while in Carolina before returning to

England in 1726. He immediately set to work on the book on Carolina. The first volume was published in 1731 and the second in 1743. The impact of this work was quite considerable and in 1733 he was elected as a fellow of the Royal Society. In 1747 he read a paper before the Society on "The Migration of Birds" in which he produced much new evidence concerning a subject which was by no means properly understood or fully acccepted at that time. He was equally learned in botanical matters and in 1737 published "Hortus Britanno-Americanus" which described several plant species hitherto unknown. He died in London in 1749.

The honour of producing the first bird book in England which contained coloured plates must go to Eleazar Albin, whose "A Natural History of Birds" was published between 1731 and 1738. This included three hundred and six plates drawn by Albin and coloured by himself and his daughter Elizabeth. In a subtitle to the book he claimed that the plates were "Engraven from the life". In fact they were extremely stiff and not at all lifelike. Albin was not an ornithologist as such, being by profession a teacher of watercolouring; but he "was drawn to the study of natural history by observing the beautiful colours of flowers and insects". The selection of birds illustrated in this book was pretty haphazard and the text of little worth: because for the first time coloured plates were included, "A Natural History of Birds" must be considered as a milestone in ornithological publishing. Albin tried hard to make his book as comprehensive as possible, inviting his readers to bring any curious birds that they might encounter to his home "near the Dog and Duck in Tottenham Court Road" in order that he might sketch them. The book was not a financial success and Albin himself remarked that his relations formed a disproportionate part of the subscribers' list!

A far more serious and important naturalist was Thomas Pennant who was born in 1726 at Downing in Flint. He was an industrious and prolific author whose books marked the end of a lean period for natural history in Britain. He is generally known not so much for his own work as for the fact that he was a major correspondent of the great Gilbert White. The book by which Thomas Pennant is best remembered is his "British Zoology". This was the first publication of any real value to ornithology since "Willughby's Ornithology". It was published between 1761 and 1766 and illustrated with one hundred and thirty-two coloured plates executed by Peter Mazel. One hundred and twenty-one plates were of birds, the remaining eleven being quadrupeds. The fifth part of this work is frequently missing, in which case only one hundred and seven plates are present. "The British Zoology" was immediately popular and encouraged Pennant to write several more books, mainly about his travels. "A Tour of Scotland" was published in 1771 followed by "A Tour of Wales" in 1778, both of which are still of great interest. His "Genera of Birds" appeared in 1773, but despite the imposing title it was one of little scientific value. Probably his most important, although not his best known, work was "Arctic Zoology" first published in 1784 with a supplement in 1787. Much of the information for this book was supplied by Sir Joseph Banks following his voyages to the Arctic.

Aside from his writings Thomas Pennant lived a busy and active life. He owned a considerable estate which he managed himself and was also a local magistrate.

He probably met Gilbert White through his brother Benjamin, who was also Pennant's publisher. Their correspondence extended over many years and Pennant published much of it without any acknowledgement as to its source — an omission which caused even the tolerant Gilbert considerable annoyance. Set against this Pennant did introduce Gilbert White to Daines Barrington, to whom some sixty-six letters on the natural history of Selborne were addressed, and who was a great deal more appreciative of White's advice and help.

During his later years Pennant was confined to his estate in Wales, owing to failing health and an injured knee. He died in 1798. He was not a great ornithologist but credit must be given to him both for his writings and his work on the taxonomic classification of birds. He corresponded extensively with Linnaeus on this subject, and it was through Linnaeus' recommendation that Pennant received the honour he valued most, membership of the Royal Society of Uppsala.

The name of Gilbert White is more closely associated with natural history than any single person before or since. It is impossible to overestimate the influence which this gentle, quiet, retiring curate, who lived for most of his life in a small Hampshire village, has wielded over more than two centuries. Countless people have been inspired by his writings to take an interest not only in natural history but in wildlife in general as it occurs in their immediate environment.

"The Natural History and Antiquities of Selborne in the County of Southampton" was first published in 1789, and since then has appeared in more than one hundred and fifty editions in many different languages. Apart from the Bible and the works of Shakespeare it must it must be one of the most frequently reprinted books in the English language. For Gilbert White no facet of life in the countryside lacked interest or was unworthy of notice. In his journals he recorded the arrival and departure of migrant birds, their periods of song, their nesting habits, the flowers and insects he saw, the state of the crops, the weather, and even the time that farmer Lassam's Dorsetshire ewes began to lamb. He set out the dates on which he planted his garden and the time his cucumbers ripened.

Events in the outside world were also noted and commented upon. The French revolution, the first crossing of the Channel by a balloon from England to France, everything was of interest to him. His letters to Daines Barrington and Thomas Pennant are full of acute observations set out clearly and concisely. In unambiguous terms he differentiates clearly between his theories and those things which he has actually seen. The unaffected simplicity and clarity of his letters and journals make it easy to appreciate not only what he observed but what he felt about it as well. So vivid is his description that a visitor coming to Selborne some two hundred years later feels he knows the place already. The zig-zag path made by Gilbert and his brother remains and time dissolves when one reaches the top of the hill and sees his beloved village stretched out below. Not that Selborne is unchanged but the feeling of the English countryside captured so perfectly in his writing is unchanging.

On a purely scientific plane Gilbert White also made many valuable contributions. He was the first to differentiate between the chiffchaff, willow warbler and wood warbler as well as giving an accurate description of the harvest mouse. His field notes on the behaviour and life histories of the birds and mammals which he observed so carefully are of inestimable value. He noted that swifts mate on the wing and that the young stay in the nest until they are fully developed and virtually indistinguishable from the adults in flight and feeding behaviour. He observed that the old nests of the house martin were invariably heavily infested with lice so as soon as the young had fledged he used to pull down the nests lest they infect later broods. These and hundreds more similar observations gave a more complete and accurate picture of the lifestyle of birds and other creatures than had ever been available.

Although the names used by Gilbert White are often different from those by which they are known today, and despite his rather erratic spelling, there are few cases in which the identity of the bird he is describing is not entirely clear. Very few mistakes occur in his observations and those that do can generally be attributed to the lack of such aids as binoculars, field guides or ringing returns.

His greatest achievement was perhaps to awaken in others the same interest and sense of inquiry into the world around them which was so strongly alive in him. A shy, diffident man he saw clearly in the natural world the hand of the God whom he served so faithfully both as a village priest and as a naturalist. He sought neither recognition nor reward and particularly requested, in an age of elaborate tombs and memorials, that he should be buried in the unfashionable part of Selborne churchyard in a grave marked by the simplest of memorial stones with only the letters G.W. 1793 carved upon it.

Gilbert White was born at the vicarage in Selborne on 21st June 1720. His father was a barrister and his grandfather, another Gilbert White, was at the time the vicar of Selborne. For a time he lived at Compton near Guildford, and then moved to East Harting. About 1730 the family moved into "The Wakes" in Selborne where Gilbert spent the rest of his life. He was educated at Basingstoke Grammar School and went up to Oriel College, Oxford in April 1740, graduating in 1743. He was ordained a Dean in April 1747 and became a curate to his uncle, the rector of Swarraton and Bradley near Alresford, at a stipend of £25 per annum. By 1751 he was back in Selborne as the curate in charge and living at the vicarage. In 1752 he returned to Oxford as a junior proctor and then dean of his college. On leaving Oxford he took a number of curacies, amongst them West Deane in Wiltshire and Durley and Newton Valence in Hampshire. From 1756 to 1759 he was once again curate at Selborne and then from 1761 until 1784 curate of Farringdon, the next parish to Selborne. The rector at Farringdon was an absentee, as was often the case in the eighteenth century, so it suited Gilbert to hold this curacy until 1784 when he finally took over the curacy of Selborne and which he held until his death. He has often quite wrongly been referred to as the vicar of Selborne, but this was impossible for him since the living of Selborne was in the gift of Magdalen College and he was an Oriel man. Throughout his life he resolutely refused all offers of promotion having no wish to leave his beloved Selborne, although he did accept the living of Moreton Pinkney in Northamptonshire, but never took up residence there. He had a total income of £300 a year, not rich even by eighteenth century standards, but sufficient for his simple needs.

Because his life was so tightly intertwined with that of Selborne it is often supposed that he never travelled beyond its immediate locality. It is true that he never went abroad or visited Scotland or Wales, but he did spend some time in Devon, went to the Fens and visited Shrewsbury, Derbyshire and Salisbury. Mainly his journeys were in Sussex and Hampshire and he regularly rode to Oxford and London. He probably went further afield than most people, even if they had the means to travel, in those days of abominable roads and no trains.

He devoted much of his spare time to improving his house and garden and made several additions to "The Wakes". This building which survives and now houses the Gilbert White Museum stands in the main street of Selborne a little way from the church. It is a large rambling old house which has been added on to greatly over the years. It is known that Gilbert White added what he called the Great Parlour at the rear of the house and greatly enlarged the garden.

From 1751 until 1768 he kept what he called his "Garden Kalender" in which he meticulously recorded everything he planted and harvested together with its progress and details of the weather. In 1768 he abandoned his Kalender for his Naturalists Journal, the diary which was to become so famous. He continued to write his Kalender but in an abbreviated form, merely giving details of wine making, beer brewing and similar subjects. The original Kalender, in Gilbert White's own handwriting, has survived and is in the British Library. A reproduction of it has been published recently and makes interesting reading.

His correspondence with Thomas Pennant was begun in 1767 and by 1780 he had written some forty-four letters to him covering a wide variety of subjects, mainly concerning natural history. To the Hon. Daines Barrington he addressed some sixty letters commencing in 1769 and the last dated 1787. A small point here nicely illustrates the endless trouble that this courteous and kindly man would go to in order to assist others. When writing to Pennant and Daines Barrington he always used the Linnean system of naming plants and birds etc. However when writing for himself in his own journals he preferred to name the birds according to Ray's classification and plants according to the sexual system.

Gilbert White's lasting earthly memorial is "The Natural History of Selborne", and also his journals, none of which was published until after his death. The originals have survived in the manuscript department of the British Museum. Following the death of his niece, Mary, in 1839, their history is uncertain but they resurfaced in 1881 when they were sold to the British Museum by the Rev. George Taylor of Pulborough in Sussex.

Apart from his connection with Gilbert White the Hon. Daines Barrington was not a naturalist of any great distinction. Born in 1727 the fourth son of Viscount Barrington, he was a lawyer of some eminence. His interests were many and varied and besides being an active member of the Society of Antiquaries, he planned an expedition to the Arctic. It was he who perceived the excellence of White's writings and persuaded him to publish "The Natural History of Selborne", and for this alone he deserves our gratitude.

The chaos into which zoological nomenclature had fallen by the beginning of the 18th century has already been touched upon. This was made increasingly worse by the flood of specimens of new species constantly being sent back to Europe and Britain from freshly discovered lands and new colonies.

Much important ornithological work was being done throughout Europe at this time, particularly in France where, unlike Britain, there was something of an intellectual renaissance. Although the achievements of Brisson, Buffon and Linnaeus lie outside the scope of this book, some reference must be made to them because of their enormous influence upon the entire zoological world.

In France the leaders of ornithological classification were Mathurin-Jacques Brisson (1723-1806), and George-Louis Leclerc de Buffon (1707-1788), who both attempted to classify and list every bird known at that time. They were well placed to do this work as both men were curators of private collections and so, between them, had access to the greatest collections in France.

Brisson's method of arranging birds was to place them, somewhat arbitrarily, in twenty-six orders as defined by beak and claws. Whilst this was a fairly flexible system it produced insoluble difficulties in some cases, and was not easy to remember. He took several of his orders directly from Ray, but did not always have examples of both sexes of some species, whilst many of the specimens described by him were in a very poor condition by the time they reached him. In all he classified some 1500 species, and some of his descriptions were extremely accurate.

George-Louis Leclerc, Comte de Buffon, was a leading natural historian during the latter part of the eighteenth century. Although his system of classification was also somewhat defective he appreciated that observations made in the wild were essential to a complete knowledge of live beings, and strongly defended the belief that all classifications should reflect the order of nature.

The failure of Brisson and Buffon to produce any lasting system of classification must be attributed to the fact that both devised narrow, rigidly structured systems which were incapable of modification or expansion.

The first truly reliable and lasting system of classification was devised by Carl von Linné, widely known as Carolus Linnaeus. He was born in 1707 in the province of Smaland in southern Sweden and studied medicine at the Universities of Lund and Uppsala. After many vicissitudes he was appointed to the position of Demonstrator of Botany, a position which he filled with great distinction. He began to build up a large collection of plants and in pursuance of this study went to Lapland, a considerable adventure at that time, involving much effort and discomfort. He also travelled to Germany and Holland and visited England in 1736. On his return to Sweden he took up medicine again before devoting himself entirely to botany. The systematic classification which he proceeded to evolve, although it has been much modified in detail over the years, is that which is still used today. His binominal system gave a name for the genus, and a second name to designate the species. This has now been expanded to include a trinominal name which denotes the sub-species or race concerned. Although not strictly speaking an ornithologist, Linnaeus fortunately included birds within his classification although the number was relatively small. Only nine hundred and thirty-one species appeared in the 12th and last edition of his "Systema Naturae Regnum Animale", and few species retain today the names given to them by Linnaeus. His importance to ornithology lies in the fact that the names he gave in the 10th edition of "Systema Naturae Regnum Animale" (published in 1758) established a firm basis from which stems all modern taxonomic nomenclature. Linnaeus' work marked the point at which zoological classification set out firmly on the path of rationality, consistency and reliability. His work had a far-reaching and lasting impact, particularly in Britain and Germany.

During the latter part of the eighteenth century North America not only obtained political independence but also began to acquire naturalists and literature of her own. Although Mark Catesby can fairly be claimed to be the first to write extensively about the birds of North America, Alexander Wilson was undoubtedly her first notable ornithologist. To some extent he has always been overshadowed by Audubon, and although always appreciated within his own country has never received the worldwide credit rightly due to him. This is partly because of the universal fame of Audubon's "Birds of America" and the huge monetary value placed upon it, and partly owing to Wilson's dour and withdrawn nature which gave a clear field to the self-confident Audubon. A great field naturalist, Wilson was studying North American bird life whilst Audubon was still painting portraits. Everything that Wilson put into his "American Ornithology" was his own work, the result of his reading and observations, or if not due acknowledgements was made. He was his own artist and wrote his own text. He saw his book through the press even obtaining the necessary subscriptions for the work by travelling far and wide on foot, horseback and boat throughout all the parts of

America that were settled at this time.

Alexander Wilson was born in Scotland in 1766 where his father after an earlier career as a smuggler, had become a weaver employing several workmen. His early childhood was a happy one, but when he was ten years old his mother died and his father's business collapsed and he was forced to take the lowly job of a cowherd. Nevertheless his love of nature and ability to endure solitude engendered at this period were to stand him in good stead during his later life. He became a weaver himself and travelled around Scotland selling silks, printed muslins, ribbons and handkerchiefs. He had written poetry since his childhood and the fame of Robbie Burns now refired his enthusiasm and he restarted writing, prose as well as poetry. However in 1792 he was imprisoned, on some trumped up and unreliable evidence, for seditious writings and blackmail. Britain was at war, the times were troubled and the authorities suspicious, and the Scottish legal system was in chaos. From 1792-1794 he was constantly in and out of prison, under licence, as the charges against him were altered variously from libel to blackmail. Under these circumstances he knew he was unlikely to get a fair trial and decided to emigrate to America, leaving Scotland on 16th May 1794 for Philadelphia.

He soon left Philadelphia to take up a post as schoolmaster in Milestown some nineteen miles outside the city, but found it difficult to fit easily into a new life and community and following an unhappy love affair resigned from the school. A melancholy character, he appears to have had some sort of mental crisis at this time as he travelled endlessly around the New York and Pennsylvania area determinedly avoiding all his former friends and colleagues. This period of wandering seems to have eased the strain he was under and he eventually returned to Philadelphia and took over a school at Gray's Ferry not far from the city. Although Wilson does not appear to have shown any particular leaning towards the study of birds whilst he lived in Scotland it would seem that he soon became fascinated by those in the New World. The number and variety of birds around Gray's Ferry increased enormously during times of migration: and Wilson became fascinated by these hordes of so many different species pouring through in spring and autumn. Becoming more and more enthusiastic he conceived, sometime in 1802, the idea of a book which would include not only illustrations of them in colour but also details of their lives. He began to sketch many live birds, shooting others in order that he might examine them more closely.

In 1804 Wilson approached Alexander Lawson, an engraver, told him of his plans for his book and enquired as to its viability. Lawson dismissed the idea, the cost of the copper plates alone would amount to $500 and the engraving more than $5,000 in addition to some $12,500 for colouring the plates. This would make the cost of the book impossibly high. Undeterred he decided to do the job entirely by himself and, borrowing some tools from Lawson, began to practise engraving. He estimated originally that he could illustrate all the birds of America in a ten-volume work, each book to cost $12 a volume and the entire set costing $120.

From late 1805 and into 1806 he worked on steadily, drawing, etching and colouring his plates. He resigned from the school and took a post as an assistant editor to a publisher who was preparing a new encyclopedia. This publisher, Samuel Bradford, agreed to publish the "American Ornithology" when it was finally completed. One of Wilson's greatest problems was with the colourists. The quality of the plates varied widely because the colours used were not uniform. Nevertheless an advertisement was prepared, together with a sample of a coloured plate showing the quality of the work, and sent to prominent people who might be interested in subscribing. Samuel Bradford undertook to underwrite the first volume for an edition of 200 copies provided Wilson obtained at least that number of subscribers.

In order to find the necessary subscribers Wilson now began travelling around America to seek subscribers for his great project. Amongst these was Thomas Jefferson, the President of the United States, who strongly approved and supported Wilson's efforts and also provided him with a further list of names of likely subscribers.

Wilson's travels took him into the southern states and west to Mississippi. Whilst these long and often difficult trips considerably taxed his strength, he saw and obtained many new bird species which added greatly to the scope and value of his book.

Throughout this time he was beset by problems. Subscriptions came in very slowly, and for a time he faltered, wondering whether the country was ready to accept a work of this magnitude. But, at last, in September 1808 the first volume was finally printed.

It is impossible to overestimate the amount of work that Wilson put into his "Ornithology". As he lacked independent means he had to work, first as a teacher then as an assistant, throughout the years it was in production.

Wilson's whole life revolved around his book. He got up before dawn, walked to his office and organised the work of the colourists, getting on with his own writing between excursions to obtain fresh specimens for the colourists to copy. Of the 343 species of birds to be found in the United States as it was constituted at that time, Alexander Wilson described 264, adding 48 new species to the total known hitherto. He gave life histories of 94 species and his errors were very few. "Science would lose little, but on the contrary would gain much, if every scrap of pre-Wilsonian writing about United States birds could be annihilated" — so wrote Elliott Coues, and surely this was the truth.

During a journey to the Mississippi region in March 1810 Wilson met J.J. Audubon, who it seems was not impressed by Wilson's melancholy manner and strange appearance. Wilson was able to show him the first two volumes of his book, and Audubon was about to order a copy but was dissuaded by his partner, Rozier, who suggested that Audubon's own bird drawings, of which he had executed very few at this time, were infinitely superior. Wilson makes few comments in his letters concerning Audubon, and Audubon's memories and remarks, all written long after the events took place were probably rather unreliable.

Towards the end of his life Wilson's relationship with his publisher, Samuel Bradford, deteriorated. Because subscribers were unwilling to pay more than $24 a year, (at $12 per volume), not more than two volumes could be published annually. Bradford became more critical of Wilson's work. The war with England in 1812 occasioned more delays in publication and therefore in the receipt of monies by Wilson.

Alexander Wilson did not live to see the completion of his great work: he died from dysentery on 23rd August 1813. By that time he had completed the plates for all nine volumes, seven of which had been published within his lifetime. The final two volumes were completed by George Ord, one of Wilson's executors, using Wilson's notes where they existed and completing the remainder from his own observations. He later reissued Vols. 7 — 9 with many changes in nomenclature and further alterations to the text. There were many subsequent editions of Wilson's "Ornithology" under different editors and publishers; including a completely new edition, in three volumes, by George Ord in 1828-29. Later on various octavo editions appeared issued by different publishers and with Charles Lucien Bonaparte, Prince of Musignano as co-author. Precise details of these editions are too numerous to list here, but the reader should consult the catalogue of the Edward Ayer Ornithological Library by John T. Zimmer, Chicago 1926, for a full bibliography.

III

The Birth Of The Modern Bird Book

During the 18th century ornithology and its literature had developed in a slow and often erratic fashion, but with the advent of the 19th century radical and far-reaching changes were about to take place. This was an age which saw a tremendous expansion both in quality and quantity in the production of books devoted to natural history.

One of the more important factors which had hindered all ornithological progress hitherto had been a lack of any kind of specialisation amongst those who seriously devoted their energies to studying natural history. There were no "specialists" in botany, ornithology or geology etc., the prevailing custom being simply to consider the natural world as a whole. Not that this attitude of mind should be entirely condemned: in these days of ultra-specialisation there are occasions when a more well-rounded and less specialised view could give a more balanced and less clinical aspect to a subject.

The early years of the 19th century saw the publication of three books which were the earliest prototypes of the handbooks with which we are so familiar today. George Montagu was the pioneer author of this type of book; through his work he considerably advanced the progress of ornithology towards becoming a separate line of study within the field of natural history.

Born at Lackham House in Wiltshire in 1751 he was one of a family of thirteen. He followed a career in the army and fought in the American War of Independence, before returning to England where he became a colonel in the county militia. Following an acrimonious lawsuit the family estates were split up and George Montagu went to live at Knowle Cottage just outside Kingsbridge in South Devon where he devoted the remainder of his life to natural history, particularly the study of birds. An energetic man and an acute observer, his contribution to ornithology has never received the recognition in Britain that it deserves. He admired and was greatly influenced by the work of Gilbert White to whom he wrote saying, "Your work produced in me fresh ardour".

Montagu's considerable field ability resulted in several important discoveries which led to the separation of species which had previously been confused. For example, the cirl bunting which until his time had always been confused with the somewhat similar yellowhammer. Montagu also identified as a separate species the harrier which bears his name and which until then had invariably been confused with the hen-harrier. He was the first person to describe scientifically the roseate tern, this he named Sterna dougallii in honour of Dr. McDougall of Glasgow who sent him the specimen, a name which it bears to this day. Besides this he recorded the first known occurrence of the little gull and the little crake in Britain.

His most important contribution to ornithology in general however, and the one for which he is best remembered is his "Ornithological Dictionary" which was published in 1802 in 2 volumes, (usually bound together). This was a really excellent book. Set out in alphabetical order it was the most complete and accurate practical handbook which had been published up until this time. Montagu's remarks on migration and bird song were very sound and in addition he listed all the known alternative names of birds of which there was such a proliferation at that time. What the work did lack was a sufficient number of plates and only one of these, that of the cirl bunting, was in colour.

The "Supplement to the Ornithological Dictionary" was published in 1813 and was, in effect, an up-to-date edition of the earlier work. The amount of new material which appeared in the Supplement was a telling indication of the amount of fresh knowledge acquired in the ten years which had elapsed between the publication of the two books. The supplement contained 24 black and white illustrations, which in a few copies were coloured. The drawings were on the whole very poorly executed, particularly with regarded to their shape. Elliott Coues writing in the USA towards the end of the 19th century referred to The Dictionary as "one of the most notable treatises on British birds, as a vade mecum which has held its place at a thousand elbows for three quarters of a century". George Montagu died of tetanus in 1817 after accidentally treading on a rusty nail.

Thomas Bewick is chiefly remembered today as one of the greatest wood engravers this country has ever produced, whose small, intricate pictures have appealed to successive generations. Whilst he was not an ornithologist in the accepted sense of the word, through his illustrations and books, he had a considerable influence on the subject.

He was born at Eltringham in Northumberland in August 1757, the son of a small tenant farmer. At the age of fourteen he was apprenticed for seven years to Ralph Beilby an engraver in Newcastle. The young Bewick was fortunate in the choice of his master, for through his association with Beilby he gained experience over a wide field of work which ranged from engraving door and coffin plates, invitation cards and clock faces to engraving sundry illustrations for other local printers. A clever, inventive man, Beilby had adapted many engraving tools to suit his own needs and these innovations were to be of lasting value to Bewick. Beilby himself was not fond of engraving on wood, and as the young Bewick showed such promise and aptitude for this type of work he was soon happy to pass most of it over to his apprentice.

On the completion of his apprenticeship Bewick returned to his home remaining there throughout the winter of 1774. During this period of comparative leisure he was able to observe closely the birds and animals of his native countryside and broaden his knowledge of the country scenes which were to provide the subject matter for so much of his later work. In fact Bewick was not a great naturalist: rather he could be described as an intelligent and interested observer whose service to natural history lay in his delightful illustrations which revealed to many of his readers for the first time the charm and warmth of the homely scenes and the beauty of the countryside which existed all around them.

After a short time in London, a city which he did not like, Bewick returned to Newcastle and set up in business on his own. However, soon afterwards Ralph Beilby offered him a partnership which, after some hesitation, he accepted. There was always a great deal of work of a general nature for him to do which meant that any illustrations which he wished to execute for his own books had been done in his own spare time.

He began his first book "The History of Quadrupeds" in 1785 and it was ultimately published in 1790 with text which was written by Beilby. The book was a great success and sold 1600 in a year. It is interesting to note that the pictures of animals which Bewick had been able to see for himself either in the wild or in a menagerie were distinctly better than those he had only been able to see stuffed in museums.

The success of "Quadrupeds" encouraged Bewick to produce a bird book which he had been contemplating for some time. Like Montagu he was greatly influenced by the writings of Gilbert White and Thomas Pennant, and he became increasingly determined to publish an ornithological book of his own. To this end he commenced work in 1791 on "A History of British Birds". The text for the first part, "Land Birds" was again written by Beilby and was of little originality or value as it was mainly copied more or less directly from the work of previous authors such as Pennant, Belon, Willughby and Ray. It was Bewick's engravings which made the book such a success; no less than six editions being published in his lifetime. The first edition appeared in 1797 and was printed on three different sizes of paper, imperial, royal and demi-octavo, and in two different thicknesses of paper. The bibliography of this popular work is, as might be deduced, complicated, and the interested reader is best referred to Mullens & Swann, "A

Bibliography of British Ornithology'' (London 1917). The second volume entitled ''Water Birds'' had its text as well as illustrations provided by Bewick. It was published in 1804 and again there are numerous editions and reprints of this volume.

It has been claimed that Bewick studied birds from his earliest youth, but it would perhaps be more accurate to say that his interest in birds was simply a result of his general interest in the natural world around him. Today he is probably best known for his tiny pictures, so fine and clear cut that they give an almost three-dimensional effect, of scenes from everyday life; such as a boy driving geese, a bull in a farmyard or a woman hanging out washing on a line.

After the completion of his bird books in 1804 Bewick did very little else directly connected with natural history before his death in 1828. No doubt, it would have given him much satisfaction if he could have known that more than a century and a half later his drawings are still in use, as vignettes in books, on note-paper and cards, and that many of his engravings are still being reproduced and enjoyed today. Bewick's engraving tools have been preserved and can still be seen at the Library in Newcastle-upon-Tyne.

The great social and industrial upheavals which took place during the latter decades of the 18th century were accompanied by equally significant changes in agriculture. Field enclosures, the discovery of the rotation of crops, the feeding of cattle throughout the winter and the selective breeding of livestock brought about changes as fundamental as those effected by the Industrial Revolution. The increase in wealth brought about by more industrialisation created a larger and more affluent middle class amongst whom were many who were anxious to use their newfound leisure in pursuit of ''improving'' hobbies and activities. This combined with the opportunity for cheap and easy travel provided by the railways resulted in the creation of numerous field clubs and societies. As Britain became increasingly urbanised attitudes towards the countryside changed and it began to be viewed in an entirely different light. Scenery which earlier had been called ''horrid'' or wild or strange was now deemed to be beautiful and romantic. There was now a need for informative books; by coincidence, new printing methods improved mechanisation and cheaper mass-produced paper meant that the demand could be met cheaply and well.

Amongst these was a book which in a variety of editions was to serve as the standard work for field ornithologists for almost the rest of the century. "A History of British Birds" by William Yarrell was first published in 1837 and completed in three volumes in 1843. It was illustrated with 520 wood engravings and 59 tailpieces. The engravings were of a high standard and were executed by John Thompson and two of his sons. The great majority of the plates were drawn by Alexander Fussell with a few being drawn as well as engraved by John Thompson. The text of "A History of British Birds" was accurate and clearly set out, and although it was published only 25 years after Montagu's "Dictionary" shows clearly in comparison how much ornithology had progressed during the intervening years. Following the fashion of the time the book was issued in 37 parts and supplements were added in 1845 and 1856. There was a second edition in 1845 and a third in 1856. The fourth edition was published after Yarrell's death and was edited by Alfred Newton and Howard Saunders.

William Yarrell (1784-1856) was a man of considerable ability and charm. He was born in Ryder Street, St. James in London, the son of a newsagent and bookseller, and was educated in Ealing. He eventually inherited his father's business which provided him not only with a reasonable income but also with the time to indulge in his favourite sports of shooting and fishing. However, his time in the countryside was not devoted entirely to sport; he also watched birds and studied the natural history of fish. A bachelor, he was an extremely sociable character as well as an enthusiastic naturalist. He was also a Fellow of the Linnean Society, one time secretary and later vice-president of the Zoological Society and a founder member of the Entomological Society, besides being a frequent contributor to all their journals. He was well acquainted with

John Gould, Prideaux John Selby and Sir William Jardine among many other naturalists.

Yarrell was well known for his friendly and helpful attitude to all who displayed an interest in natural history. He was particularly happy at being able to meet Thomas Bewick when towards the end of the engraver's life he paid a visit to London, to discuss birds and fish with him, and show Bewick his private museum containing many stuffed birds and objects of interest to the naturalist. He always held Bewick in the highest regard and was instrumental in naming Bewick's swan in his honour. Unfortunately this act was to cause much controversy and ill-will in some quarters. John Hancock, the founder of the museum which bears his name in Newcastle-upon-Tyne and R.B. Wingate, an expert taxidermist in Newcastle had earlier come to the conclusion that this swan, eventually to be named Bewick's swan, was a different species from the whooper swan with which it had formerly been confused. However they did not publish their finding; when Yarrell did so they were incensed. Their anger and bitterness was the more intense because they felt that as Geordies and "provincials" their work was not taken seriously in London.

This unfortunate incident does not appear to have been prompted by any malice on Yarrell's part, in fact it would seem that he hesitated for about four years before he formally named the swan. It was an unhappy episode and one which achieved some prominence in "The Philosophical Magazine" which published numerous letters from its readers taking sides over the rights and wrongs of the matter. Further confusion was caused by the fact that many considered that it was Pennant who first identified the bird and Yarrell who named it None of this controversy and petty squabbling detracts in any way from the status of Yarrell's work: "A History of British Birds" was a milestone in ornithological publishing which has become a classic.

A parallel work by a contemporary of Yarrell's, William MacGillivray, although in many respects a valuable book, was by comparison a failure. The reason for this probably lay more in the nature of the author's character than any lack of ability. MacGillivray was a highly sensitive, touchy man who in spite of possessing considerable ability and intelligence appears to have suffered from a strong sense of inferiority. He felt, with some justification, that he was neglected by the public, especially in England. He dreaded failure, but the harder he tried to achieve success the more irritable and difficult he became. His openly expressed contempt for all "compilers and theorisers" and his utter disregard of those he considered to be fools brought him into conflict with many people. It must be admitted that his peculiar classification of birds was not only confusing but definitely a retrograde step at a time when ornithological nomenclature was progressing towards some uniformity. For instance, there was no justification whatsoever for calling the blackbird the black thrush, or the wheatear the white-rumped stonechat, as well as other highly unusual names. MacGillivray's book "A History of British Birds — Indigenous and Migratory" was published in 1837, the same year as Yarrell's book and bearing much the same title. The fact that it had far fewer illustrations than its rival was probably another factor which weighed against it in the eyes of potential buyers. In some respects it was the better book in that it contained a great deal of original and accurate material. In all it took fifteen years to produce the five volumes which comprise the entire work, being completed just before MacGillivray's death in 1852. An unillustrated pocket edition "Manual of British Birds" was published in 1840-42. In 1836 he published "Descriptions of the Rapacious Birds of Great Britain" and in 1839 a book on mammals entitled "British Quadrupeds".

However MacGillivray's greatest contribution to ornithological literature, certainly the one for which he is best known, lies in the part he played in Audubon's great book "Birds of America"; probably the most famous, and certainly the most expensive bird book, ever to be published.

The first part of the first edition (which was printed in

Edinburgh) consisted solely of plates, the reason being that if the book had contained any text Audubon, under the British copyright law as it existed at that time, would have to deposit a total of nine copies in specified libraries. This he could not afford to do and instead published the five volumes of text separately under the title of the "Ornithological Biography, or an account of the habits of the birds of the United States of America" (1831-39).

At this time Audubon was forced to devote all his energies not only to finding subscribers for his book but to supervising the printing and production of it. Besides lacking the time he soon realised that he also lacked the necessary knowledge to write the taxonomic section and the anatomical detail. Audubon wrote of his problem as follows "I applied to Mr. James Watson, to ask if he knew of any person who would undertake to correct my ungrammatical manuscripts, and to assist me in arranging the more scientific part of the biography of birds. He gave me a card with the address of Mr. MacGillivray, spoke well of him and his talents, and away to Mr. MacGillivray I went. He had long known of me as a naturalist. I made known my business, and a bargain was soon struck. He agreed to assist me and correct manuscripts ...... and that day began to write the first volume".

MacGillivray also assisted Audubon by writing much of the text for "A Synopsis of the Birds of North America" which was published in 1839. This book was primarily a classification with notes concerning birds already included in "The Birds of America" together with a number of small revisions of the earlier work.

Of the man and his work we know less than we would wish, almost all his notes and journals having been destroyed in a fire soon after his death. From references made about him by others we know something of his general character but very little about his thoughts or mental processes. He was born in Aberdeen in 1796 but spent most of his childhood in the Isle of Harris in the Hebrides before returning to Aberdeen in 1808 in order to enter the University there. He studied medicine and in

1815 was employed as Dissector to the Lecturer of Anatomy, but he later forsook medicine to study natural science.

During this period of his life MacGillivray travelled widely around Scotland and being very impecunious at this time set out to travel to London on foot. He later recorded that during this arduous trip he lived mainly on bread and water having only 12/- (60p) on which to feed himself whilst covering the final 240 miles. He did not like London and he did not like the English nor they, it would seem, him: so despite all his efforts the expedition was rather a disaster.

In 1820 he went to Edinburgh and in 1823 became assistant to Professor Jameson, the Regius Professor of Natural History. From 1831 until 1841 he was the conservator of the museum of the Edinburgh College of Surgeons, and it was during this period that much of his best work in the field of natural history was accomplished.

It was unfortunate that owing to the pressure of other work and his hypercritical perfectionist attitude towards his own work there were such delays in the completion of "A History of British Birds". The hiatus which occurred between the publication of each volume undoubtedly lost MacGillivray many sales, with the passage of time purchasers tended to lose interest. These intervals between volumes, especially when compared to Yarrell's work, with whom in a sense he was competing for the same market, were greatly to the latter's advantage.

Although technically he was very capable, William MacGillivray's anatomical work was somewhat uneven. As with his nomenclature of birds he was inclined to go off at a tangent and miss the important salient fact. He came very close to discovering the taxonomic importance of the musculature of the vocal organs, but instead veered off from this line of research in order to concoct a classification based upon the structure of the digestive organs, which was virtually useless.

On leaving Edinburgh MacGillivray returned again to Aberdeen where he was appointed to the professorship of "Civil and Natural History" in Marischal College. Apart from his academic work he undertook a number of journeys during which he observed at first hand the wild life around him and also visited various museums in order to make notes on their collections and administrations. As an example of the energy of the man, in 1833 within three weeks and four days he visited museums in Glasgow, Liverpool, Dublin, Bristol and London and wrote detailed reports on all of them.

Soon after 1850 he became ill and was advised to move to a less rigorous climate. Accordingly in the autumn of 1851 he moved to Torquay in Devonshire on the south coast. He returned to Aberdeen in 1852 and died there in September of the same year. In the preface of the last volume of "A History of Birds" he wrote — "Commenced in hope, and carried on with zeal, though ended in sorrow and sickness, I can look upon my work without much regard to the opinions which contemporary writers may form of it, assured that what is useful in it will not be forgotten.... and knowing that it will powerfully influence the next generation of our home ornithologists". In a way he was right, for although his own works were largely ignored by his own generation, yet through his association with Audubon, many have come to appreciate not only his scientific ability but his clarity and accuracy as a writer; and hold him in a greater esteem than was ever accorded to him by his contemporaries. The work of Montagu, Yarrell and MacGillivray had set ornithological literature on a secure and scientific basis which proved to be of inestimable benefit to those who followed them.

IV

Audubon and Gould

The books Montagu, Yarrell and MacGillivray had written were essentially practical reference works providing reliable details and descriptions of the birds, presented in an easily accessible form. Being, for the most part, somewhat sparsely illustrated, these were not field guides as we know them today, but they could be easily consulted and were of a reasonable size.

During the 19th century there emerged a whole series of magnificent books which were almost as much works of art as ornithological works. The time and effort involved in writing, illustrating and publishing such tomes virtually forced their authors to become in effect full-time professionals. Since ornithology had not yet attained a professional status, perhaps a better description for these people would be ornithologists-publishers, amongst whom John James Audubon and John Gould were the major examples.

John James Audubon's life and work has been as much distorted by the efforts of later writers to present him in a glamorous light as by his own fabrications and half truths. Whilst none of these fantasies can in any way detract from the greatness of his work, they do make an unbiased appraisal of it somewhat difficult. The evolution of "The Birds of America" became so closely entwined with Audubon's own life that the story of one becomes the biography of the other.

John James Audubon was born in San Domingo in 1785 the illegitimate son of Jean Audubon, a French naval officer turned sugar plantation owner, and Mlle. Rabin, a creole girl. There would appear to be little justification for Audubon's claim that he was in reality the ill-fated Dauphin of France, son of Louis XVI. Audubon's father openly acknowledged his two illegitimate children by Mlle. Rabin, taking them with him when he returned to France in 1789 shortly after the beginning of the French Revolution, a cause which he strongly espoused. The family settled at Couron, five miles from Nantes. The young Audubon's education was rather sporadic, he was never able to write grammatically in either French or English despite the untiring efforts of his stepmother, Anne Moynet Audubon, who was genuinely devoted to him and truly grieved when he set off for America in 1803. There he settled for a time at Mill Grove the family home in Pennsylvania but did not fit in or enjoy the life there at all and soon quarelled with many people including the man appointed by his father to look after his interests.

He returned to France in 1805 where he passed his time in shooting and hunting — his favourite pastimes — but is also believed to have studied in Paris at the Atelier of the famous artist Jacques Louis David.

In 1806 he returned to the United States with Ferdinand Rozier with whom he was later to enter into a business partnership. Again Audubon soon tired of the life at Mill Grove and following his marriage to Lucy Bakewell, the daughter of a near neighbour, he and his wife moved to Louisville in Kentucky. Here he and Rozier opened a store. The business did not prosper and the partnership was dissolved in 1811 and the Audubons moved to Henderson where Audubon scraped out a precarious existence by painting portraits.

After being declared a bankrupt in 1819 he travelled to Cincinnati where he worked for almost year in a private natural history museum under the direction of a Dr. Daniel Drake. The short period that he spent at this museum was to be of inestimable value to him in later life since now he learnt something of the art of taxidermy and also a little about taxonomy. In 1820 he began his journeyings along the rivers to New Orleans, collecting specimens and drawing as he went. By 1824 he was back in Philadelphia trying to find a publisher for his proposed book. He was not successful, and after returning to Lousiana for a while in order to raise the money for the fare, he sailed for England in May 1826.

On his arrival he immediately commenced on his search for subscribers, travelling to Liverpool, Manchester and then to Edinburgh, where he approached the well known engraver William Lizars. Much impressed by Audubon's work Lizars eagerly agreed to engrave and print 400 plates (there were eventually to be 435). Boldly, Audubon had decided that his birds must be illustrated life size in order to do them justice; and he therefore had to select an extremely large size of paper, double-elephant folio which measured 27" × 40".

In order to obtain subscribers for this ambitious project Lizars determined that Audubon must publicise himself and his work as widely as possible. Not that Audubon needed any lessons in self-promotion. He was a great social success in Edinburgh where he was feted and lionised. He liked to present himself as a simple "American woodsman" cultivating this image carefully with shoulder length hair and strange outlandish clothes. He gave numerous lectures about his travels which were simply and attractively presented although perhaps not always strictly truthful!

Once he had finally agreed terms with Lizars Audubon set off again in search of fresh subscribers. On arriving at London he received the disturbing news from Lizars that his colourist had gone on strike and he would therefore be unable to fulfil the orders. Audubon had by now begun to supply his subscribers and could not afford to disappoint them and there now ensued a frantic search for a new engraver-printer until by the greatest good chance he came upon Robert Havell in London. He could not have found a more suitable man for Havell was not only a superb engraver and artist but was also completely reliable and efficient with regard to delivery and completion dates. There is no doubt that in many cases it was Havell's skill and artistry which provided the highly finished backgrounds for Audubon's marvellously detailed bird paintings. Anxious that the pattern of the plumage of the birds should be absolutely accurate Audubon invariably used recently killed specimens as models rather than relying upon stuffed examples, constructing a framework of wires in order to place the bird in the correct position for his drawing. Whilst it is possible to criticise his painting on scientific grounds regarding their strained posture and strange position, their artistic excellence cannot be denied.

Audubon's insistence on the use of recently killed birds meant that over the years he was responsible for the death of some thousands of birds; whilst this was by no means considered reprehensible at the time, it does seem a little ironic that his name today is synonymous with the greatest and most powerful conservation society in the world!

In view of the fantastic prices realised today for Audubon's plates it is of some interest to note that the cost of 100 parts consisting of five plates each was as follows:- engraving £42, lettering £42. 10s. 9d (£42.54p), paper £14. 1s 0d. (£14.05p), printing £11. 5s. 0d (£11.25p), plus 100 tin cases to hold the plates for carriage at £5.00. In all a cost of £114. 16s. 0d. (£114.80p) against a sale price of £2. 2s. 0d. (£2.10p) per part totalling altogether £210.00. This showed Audubon a nominal profit of £96. 4s. 0d. (£96.20p) per part.

The amount of work required from Audubon to find subscribers for his work and at the same time to arrange and supervise the printing and distribution was prodigious: added to this there was the necessity to paint pictures for sale in order to provide himself with funds on which to live and pay the printers until the subscription monies came to hand. Several subscribers failed to keep up with their payments, and this required additional time and trouble on Audubon's part to find replacements for them.

In April of 1829 he returned to the United States in order to effect a reconciliation with his long suffering wife, Lucy. He also took this opportunity to gather more material to add to the species in his book and draw revised plates of some of the others.

In the spring of 1830 he returned with Lucy to England and immediately went to Edinburgh where he was introduced to

William MacGillivray with whom he planned and wrote the text section of his book which was published separately as "The Ornithological Biography" in 5 volumes between 1831 and 1839.

The writing of this book, in addition to his other work, imposed a great burden on Audubon who later recorded that he worked regularly from four in the morning until ten or eleven o'clock at night. William MacGillivray worked similarly long hours and between them the first volume was completed within three months.

At this point Audubon decided to return yet again to the United States to organise exhibitions of his work, and drum up more subscribers. He stayed for three years in North America during which time he explored Florida and the Keys travelling as far west as Galveston in Texas and Fort Union in North Dakota, but never realising his ambition to visit California and see the Pacific Ocean.

He returned to England for the last time in 1837 and immediately had to make the decision to extend the work beyond the eighty parts which he had originally advertised in his prospectus, and to increase the number of plates from 400 to 435. During the course of the production of his book many new species had been discovered and naturally he was reluctant to omit them. A few subscribers objected to having to pay for more parts, but 161 stayed the course, 82 in the United States and 79 in Europe. So in June 1838 the final volume of this mighty work was completed, having cost Audubon in total the then astronomical sum of £28,910 to produce.

In May 1839 the fifth and final volume of "The Ornithological Biography" was published and Audubon's work in Europe was finished. He returned to America in the summer and straight away commenced to work on a smaller edition, in octavo format, of his "Birds of America", besides producing "The Viviparous Quadrupeds of North America", a publication which falls outside the scope of this book.

Between 1840 and 1844 a fresh edition (octavo size) of "The Birds of America" was successfully issued. It was divided into

100 parts at a cost of one dollar per part. Each part contained five plates and seventeen new species were included. The work was designed to be bound in seven volumes when complete. Although it was still an expensive book, this was a successful venture, and in its revised smaller format ran to some nine editions, the last being in 1871. As far as is known there have been no further editions of the octavo sized "The Birds of America" until a reprint by Dover Publications, Inc. (New York) in 1967. The completion of this work marked the end of Audubon's ornithological publications. Although he continued to travel widely, going as far as the Upper Missouri and Yellowstone rivers, he produced little if anything of further value. By 1847 his mind had begun to fail and he could no longer draw or paint. He died, a sad shell of his former self, on 27th January 1851.

He was a complete paradox: vain, capable of many untruths, but an accurate and reliable observer of the natural scene; full of charm, but frequently spiteful; a superb artist and wonderful in his sense of composition, but scientifically inaccurate in the physical shape of his subjects; a competent business man so far as his publications were concerned, but a hopeless bankrupt in his commercial enterprises.

To summarize his work it is hard to better the comment of Sacheverell Sitwell who wrote,""The Birds of America" is a heroic undertaking, and that one man should have endured the hardships and ardours of so many long and lonely years, painted the pictures, written the text, and continued the publication upon so gigantic a scale, puts him among the immortals. Audubon's "Birds of America" is one of the most valuable books in the world.... Audubon is the greatest of bird painters".

Perhaps Audubon did not fully understand himself, "I am not a learned naturalist, I am only, and that to a very great extent, a practical one". This was not by any means the whole truth for he had read widely in his time and learned much from MacGillivray about taxonomy and ornithology in general. Whatever else, Audubon was not a simple man.

As for the great double elephant folio, the exact number of complete sets extant is not known for sure, certainly there are less than the 200 originally printed. Waldemar H. Fries, the greatest authority on the elephant folio estimates it as somewhere around 175. Some sets were left incomplete because the subscriptions were allowed to lapse, a few were broken up for the plates, at least ten are known to have been destroyed by fire, and some fourteen sets, which were known to exist at some stage, could not be traced by Waldemar H. Fries despite the most rigorous research.

"The Birds of America" is probably the most famous bird book ever to be published, and certainly the most valuable, a complete set having been sold in London for £1,100,000. The prices of sets broken for the individual plates has exceeded $1,500,000 and a complete work is known to have been sold by private treaty for over $2,000,000.

Although he never achieved the fame and distinction accorded to Audubon, a contemporary of his, Prideaux John Selby should not be ignored. His "Illustrations of British Ornithology" published in 19 parts between 1821 and 1834 was, quite literally, a grand book. It was printed in two volumes on 27"×22" elephant folio paper which again enabled the birds to be drawn to full life size. In many ways Selby could be said to be the English equivalent of Audubon. Like Audubon his book (the text of which he wrote entirely himself in two volumes) was published separately and in a smaller format than the plates.

"Illustrations of British Ornithology" is a difficult book to collate since the number of plates vary in different copies between 218 to 222 and some uncoloured copies are known to exist. Selby was his own artist producing around 200 of the plates, the remainder being executed by his brother-in-law, Admiral Mitford. The colouring of the plates is particularly beautiful and accurate, the birds depicted are altogether more lifelike in their shape and posture than many of Audubon's. Selby encountered great difficulties in finding a suitably competent publisher and finally engraved many of the plates himself.

Prideaux John Selby was born at Alnwick in Northumberland in July 1788 and went up to University College, Oxford in 1806, but left without taking his degree before succeeding to the family estate at Twizell (Northumberland). Apart from his great book he contributed much material to Sir William Jardine's Naturalist Library series writing the text for the two volumes devoted to parrots and pigeons; he was also one of the editors of the Magazine of Zoology and Botany. He took an active part in the local history society, and in 1833 made a lengthy trip around Sutherland in Scotland in order to study in detail the natural history of that region. He died in March 1867. A complete set of his "Illustrations of British Ornithology" would be extremely difficult to obtain today, and if found would fetch something in the region of £10,000.

The greatest publisher of ornithological literature in Britain in the 19th century was undoubtedly John Gould. It is tempting but unwise to compare him with Audubon since there were many differences both in circumstances and outlook between the two men. A fair assessment of them might to be say that Audubon was a supreme bird artist and Gould a very fine bird illustrator.

That said, it is extremely difficult to assess with any accuracy Gould's artistic ability since most of the drawings and all the engravings in his book were executed by others, namely Elizabeth Gould, his wife, John Wolf, Edward Lear, H.C. Richter and W. Hart. However the rough sketches and designs for each plate show that he did have an appreciable artistic ability. Above all it was his imagination, his business efficiency as well as his excellent choice of artist which have made "Gould's Birds" so famous and sought after.

Born in September 1804 at Lyme Regis in Dorset his family soon moved to Guildford before his father became a gardener at Windsor Castle. At the age of fourteen John Gould also entered the royal service as a gardener's boy, one of his earliest tasks being that of collecting dandelions for Queen Charlotte's tea. He soon began to trap and stuff birds and it was not long before he became a very competent taxidermist, an accomplishment which was to be of great value to him in later years. After a short period working as a gardener at Ripley Castle in Yorkshire, at the age of twenty, he went to work at the Zoological Society in London as a taxidermist where in 1827 he was appointed chief taxidermist and curator of birds.

In 1829 he married Elizabeth Coxen, who also happened to be a talented artist. Soon afterwards, quite fortuitously, he obtained a large collection of Himalayan bird skins and decided to produce a book illustrating them. In this project Gould immediately displayed that quality at which he was so adept — that of finding the right people to do the work which he knew to be required but for which he lacked the necessary skills.

Gould made all the rough sketches himself, he then persuaded his wife (who had no experience in this sort of thing) that she could transfer the drawings onto lithographic stones. Fortunately, she was able to rise successfully to this challenge. Realising that he did not have the necessary scientific knowledge to describe correctly new bird species Gould persuaded Nicholas Aylward Vigors, who was the secretary of the Zoological Society from 1826 to 1832, to describe the birds for him. This however was the only time in which Gould did not write the text for one of his own publications.

Gould then tried to find a publisher but was unable to find anyone willing to risk publishing the grand and expensive folio size bird books that he envisaged. Therefore, like Audubon before him, he was forced to become his own publisher and also to find people who were prepared to subscribe to his work. The first of his great folio books "A Century of Birds" was issued in 20 monthly parts between 1831-1833 and contained 80 plates. This was followed by "The Birds of Europe" 1832-1837 with 448 plates and "A Monograph of the Ramphastidae, or Family of Toucans" 1832-1835 with 33 plates. In these works

Mrs. Gould received assistance with many of the plates from Edward Lear.

John Gould's publications were large and numerous, (some 49 folio volumes alone). An excellent bibliography is available in "John Gould — the Bird Man" by Gordon C. Sauer, London 1982.

Gould's entrepreneurial energy and drive was phenomenal, and he succeeded in keeping up the momentum of his publications, and finding buyers for them until the end of his life.

Always in search of fresh material he also travelled widely, spending two years in Australia as well as making short visits to Paris, Switzerland, the south of France, Germany and Norway, and later on went to the U.S.A. and Canada. Without doubt his greatest ornithological achievement was his pioneering exploration and collecting journey around Australia. He set out on the four months' voyage in May 1838 taking his wife and an assistant, John Gilbert, with him. The expedition started in Tasmania and thence to New South Wales and South Australia. John Gilbert travelled on alone to Western Australia and stayed on there until 1841. He returned to England for a short while and then went back to Australia again in 1842 and visited every state there before being killed by a native in 1845. Between them he and John Gould collected some 600 species of which 300 were new to science although later research has determined that many were not in fact full species, but subspecies.

During this time in Australia Gould's capacity for hard work, always remarkable, was quite prodigious. He sketched the birds and made notes on their food, habits etc., besides supervising his wife's drawings. He kept no journal during this time so to some extent we have to trace the course of his travels by the places from which he recorded the birds. Undoubtedly a large proportion of the work of collecting specimens was the work of Gilbert which is scarcely surprising considering the range of Gould's activities during this period. Elizabeth Gould was also kept extremely busy working at producing highly finished drawings from Gould's rough sketches besides giving birth to a son (one of eight children) whilst in Hobart. The Goulds returned to England in April 1840 arriving in September of that year: by December Gould had issued the first part of his "Birds of Australia".

In August 1841 Elizabeth Gould died of puerperal fever following the birth of her last child. It was a terrible blow for Gould not only from a personal but also from a business point of view. He was in the midst of publishing "The Birds of Australia" as well as "A Monograph of the Macropodidae, or family of Kangaroos", and any delay in publication at this stage could only result in lost subscriptions. The situation was saved by his discovery of Henry Constantine Richter who was a very fine artist indeed. Richter's life appears to be a complete mystery and absolutely nothing would seem to be known about him except that he died from influenza in March 1902 at the age (or so it was believed) of 81.

The situation thus saved, the work continued; Richter then went on to draw plates for "A Monograph of the Odontophorinae, or Partridges of America", followed by one of Gould's most successful works, "Monograph of the Trochilidae, or family of humming-birds" which was published between 1849 and 1861. This last work contained 360 plates although in this case Richter was helped by William Hart who did the pattern plates and some of the colouring. William Hart also drew a number of plates for other works of Gould's, and the two men would seem to have shared a typically Victorian taste for strong bright colours. Gould would dearly have liked to have employed Joseph Wolf, a particularly fine bird artist, on a regular basis. Wolf did indeed draw a few plates for him but did not appear to like Gould personally, decribing him once as "The most uncouth man I ever knew".

Astute businessman that he was, Gould was most anxious to publish a book on British birds, knowing that it was likely to prove very popular. "The Birds of Great Britain" was published in 25 parts between 1862 and 1873, all the plates being the work of Hart and Richter after John Gould had sketched the layout etc. These plates were some of the most attractive which Gould ever produced, and in many instances included illustrations of the young birds as well as the eggs. However it did not meet with univeral approval and the great Professor Alfred Newton writing to his brother Edward said, "By the way, Gould always sends me his "Birds of Great Britain to look over for him, and the utter ignorance they sometimes betray is amazing. He has no personal knowledge of any English birds, except those found between Eton and Maidenhead, and about these species he fancies no one else knows anything. It is most amusing to see how anxious he is to avoid committing himself about Darwin's theory, of course, he does not care a rap whether it is true or not — but he is dreadfully afraid that by prematurely espousing it he might lose some subscribers, though he acknowledged to me the other day he thought it would be generally adopted before long".

The obituary of John Gould in "The Ibis" which is unsigned, but may be the work of Osbert Salvin or Philip Sclater (the editors), is not particularly complimentary. "There is too much of the studied effect, which detracts from their scientific accuracy. To a patient ornithologist, seeking only the correct determination of a species, a reference to one of Gould's plates is not always so satisfactory as could be desired". On the other hand Lord Lilford was more complimentary when he wrote, "For really beautiful and correct illustrations of British birds you will find Gould's great work on that special subject in the library, but the books are so large that you would require a boy to help you carry them from the house".

After the publication of "The Birds of Great Britain", Richter ceased to work with Gould, William Hart took over completely the task of providing the finished drawings and produced some excellent results.

Gould was for ever planning new works and took tremendous pains to obtain new specimens to include in his books. He searched constantly through dealer's stocks for skins and specimens and visited museums throughout Europe in order to see and sketch new species. He was an inveterate "borrower" of specimens from his friends and acquaintances and was reputed to be not too overzealous in returning such loans. Much of his great collection, (some 6,315 skins) was eventually purchased by the British Museum (Natural History).

During his last years Gould's health became increasingly poor, but despite this, and the fact that during his last few months he was confined to his sofa he continued to work on. He planned and published "The Birds of New Guinea" and the "Monograph of the Pittidae" all of which were left incomplete at the time of his death in 1881. Eventually it fell to Richard Bowdler Sharpe to complete the work on "The Birds of Asia" and "Birds of New Guinea".

The sheer scale of John Gould's work makes it difficult to evaluate its worth. He was an outstanding publisher, business man, taxidermist and salesman. As time went by he undoubtedly did acquire considerable ornithological knowledge which was more of the museum that the field variety. Whilst not a very good field naturalist, it would seem that he possessed a great ability to recognise differences between specimens and their relationship to one another. His artistic ability is even more difficult to assess as beyond layout sketches and instructions he relied so heavily on the work of others. It is impossible in many cases to determine how much of the work on the plates was his and how much the work of others. A difficulty made worse by the fact that once he had paid for a drawing Gould considered that he was quite entitled to add his name to the signature of the artist and engraver.

As regards the contribution made to art by John Gould one can hardly do better than quote the conclusions of Dr. Allan McEvey, who has made an exhaustive and detailed study of Gould's work over many years. In "John Gould's Contribution to British Art" (Sydney 1973) he writes, "Gould was not a practising artist in the sense in which that term applied to Wolf, or even to Keulemans. Nevertheless this particular reflection of his age in the fine print of ornithology, a grand expression of nineteenth century wonder at the richness of birdlife in the Empire and exotic lands, and, finally, a portrayal of its peculiar British beauty were ultimately his.

These, initially through the vitality of his rough sketches and then his sheer tenacity of purpose, were John Gould's contribution to British art''.

V

Other Early Professionals

Among the many outstanding ornithologist who worked in the nieteenth century few would dispute the pre-eminence of Alfred Newton. His greatness lay not only in the breadth of his knowledge but also in the leadership and guidance he provided for others. Although the nineteenth century, particularly the second half, abounded with men many of whom made considerable contributions to British ornithology, they were for the most part gifted and enthusiastic amateurs. Being frequently possessed of appreciable financial means they were often highly individualistic. Against this background of numerous amateurs "doing their own thing", the knowledge, fixity of purpose and above all, discipline, of Alfred Newton made him the doyen of British ornithologists.

Probably his greatest achievements were "The Dictionary of Birds", and his part in the formation of the British Ornithologists Union together with its learned journal "The Ibis". He was professor of Zoology at Cambridge for forty-one years. He obtained this position because of his knowledge of zoology in general and anatomy and osteology in particular, and not in any way because of his ornithological achievements, ornithology being barely considered as a serious subject, let alone a science.

Alfred Newton was born in Geneva in June 1829. His father, William Newton, owned the great sporting estate of Elveden near Thetford in Norfolk. The richness of the bird life in this area must have contributed to his early interest in natural history. As a small child he had a very serious accident which left him lame, and this disability had a limiting and lasting effect on his work in the field. This comparative immobility may however have contributed to the thoroughness of his written work. He was educated privately before going up to Magdalene College, Cambridge in 1848. Here he gained the reputation of being an excellent essayist, although he, personally, had a rather low opinion of the curriculum of the University in general. In 1853 he was awarded the Drue Drury Travelling Fellowship for the "sons of Norfolk gentlemen", and often used to remark on how lucky it was for him that Elveden was just inside the Norfolk border.

He spent the following years acquiring knowledge and experience abroad. In 1855 he went to Lapland with his close friend, John Wolley, and in 1858 the two of them went to Iceland in order to make a pilgrimage to the last known home of the great auk. This bird fascinated him and during a cruise around the Northern Isles in a friend's yacht he was able to explore a rock platform near Papa Westray in the Orkneys where as he remarked, "There was enough room for a regiment of auks to have landed at any state of the tide and to have marched in line up the gentle crescent". In 1870 he made a catalogue of the remains of the great auk in as far as they were known to him. His great interest in the Garefowl, as he preferred to call this bird, was also shared with John Wolley, who was primarily an oologist.

Another bird in which Newton took a great interest was Pallas' sandgrouse, a bird which on occasions irrupts from its home in the steppes of Central Asia and has in the past reached Britain in considerable numbers, a few pairs even having bred here. The causes of these irregular visits by these birds particularly aroused Newton's interest, and he contributed a long paper on the irruption of 1863. It gave him particular pleasure to be able to exhibit a chick which had been hatched in Morayshire at a meeting of the British Association.

Alfred Newton's last expedition to the north was to Spitsbergen in 1864 with Sir Edmund Birkbeck. In 1857 he travelled to the West Indies and the U.S.A., and in 1862 he visited Madeira. In 1866 he was elected to the Cambridge Chair of Zoology and Comparative Anatomy.

His main literary contribution to ornithology was as the co-editor, with Hans Gadow, as well as being a major contributor, to "The Dictionary of Birds", published in four parts between 1893 and 1896. This was an encyclopedia of ornithological knowledge as far as it was known at that period, this clearly revealed how far the subject had progressed since Montagu's time. Professor Newton also edited the first two volumes of the fourth edition of Yarrell's "A History of British Birds" a greatly enlarged and almost rewritten edition of this important book in the preface of which Newton wrote laconically, "I am not responsible for anything that may follow by another editor", A.N.. Since the very able Howard Saunders was the editor of the other two volumes he need not have been unduly concerned!

It had originally been intended that he should revise all four volumes but he worked so slowly and meticulously that it took him ten years to revise just two of them and at that stage the publishers threatened legal action. This edict naturally exacerbated Newton's dislike of all publishers already incited by the violent quarrels which he had had with his publishers (A. & C. Black) when he was working on "The Dictionary of Birds". "Publishers I really believe are all scoundrels alike, especially those of the highest repute", he wrote; and added, "One must be dumb before the shearers because one can't help oneself. They keep well within the law, which it is their business to know, but the law enables them to fleece their victims at pleasure".

A noteworthy feature of Newton's life were his "Sunday evenings". These were informal meetings to which a great many of his Cambridge friends and acquaintances were invited. These meetings were held in his rooms at Magdalene College, in a room described by one who often attended as a "dismal room which in its spaciousness and lack of comfort resembled a barn". Here, wearing an alpaca jacket and with his waistcoat left unbuttoned, the professor would sit facing his guests who sat around him in a semicircle on hard and uncomfortable chairs. The conversation would range far and wide on many subjects and beer, tobacco and coffee were provided. He invited undergraduates to his evenings as freely as the more eminent members of the University: ladies however were not included since, although invariably most courteous towards them, he disliked and slightly feared women. No female eye was allowed to gaze upon his museum or his magnificent collection of eggs. Along with his encouraging and kindly nature he possessed considerable charm, and when discussing him with a friend who as a student some fifty years previously had known him well, it was clear how deep was the regard and affection in which the professor was held.

He could, however, when necessary, be sharply critical, and some of his book reviews or comments on the work of others were often more than a little acid. When Henry Seebohm attempted a new classification of birds, a subject about which he knew very little, Newton dismissed him as "A sociologist posing as a scholar". But to his friends he was always loyal and generous. For example when his name was put forward for election to the Royal Society, an honour which he greatly coveted and valued, he asked that his name be withdrawn in order that his friend, Canon Tristram, who was very ill at the time, be elected in his place.

His efforts in promoting the Protection of Birds Act was considerable, and his position in the scientific establishment was such that his opinions carried considerable weight. He was appalled by the actions of men such as Edward Booth, Charles St. John and John Colquhoun who prided themselves on slaughtering almost every rare bird that they encountered. Species such as the golden and white-tailed eagles were shot in horrifying numbers. The trade in feather for ladies' hats at this time meant the wholesale slaughter of such birds as egrets and grebes. For many years the campaign against this trade had been led by the Duchess of Portland and Mrs. Margaretta Lemon to whom Alfred Newton gave his continuous support. The final enactment of the Bird Protection Act of 1869 owed much to his influence and efforts.

Following his friend John Wolley's death in 1859 his great

interest in birds' eggs led him to complete the "Ootheca Wolleyana". This truly monumental work on eggs was a considerable achievement taking from 1864 until 1907. Working closely from Wolley's original notes, nevertheless he scrupulously checked everything he wrote and his tardy progress led yet again to a running battle with the publisher, R.H. Porter.

Alfred Newton maintained his interest and influence on British ornithology until his death in 1907. He finished in splendid style: it is recorded that as the end drew near the Master was sent for and prayers said, the Professor wished him goodbye saying, "God bless all my friends, God bless the College and may zoology continue to flourish in the University". When finally he could barely speak he asked to be placed in his chair for, he said, "I must die in my chair like dear Bradshaw" (A previous Regius Professor).

When Newton was awarded the Gold Medal of the Royal Society in 1900 Lord Lister summarised his work thus "The progress of ornithology in this country is due mainly to his critical, suggestive and stimulating influence". A statement with which few would have disagreed at that time or since.

A characteristic which many Victorians seemed to have in common was an absolute superabundance of energy. From the Prince Consort downwards the sheer amount of work, both mental and physical, accomplished at all levels of society was often astounding. To the Victorians to be idle and not usefully employed and failing to use one's time to the absolute limit was looked upon as little less than sinful, and this would seem to be especially true of the many writers and naturalists throughout this period. The biographer of the Rev. J.G. Wood, (one of a band of popular writers who literally poured out books and articles on a whole variety of subjects during the middle years of the century), states that his subject's "power of work was simply astonishing.... he was always at his desk by half past four or five o'clock in the morning at all seasons of the year, lighting his own fire in the winter, and then writing steadily until eight o'clock. Then in all weathers he would start off for a run of three miles over a stretch of particularly hilly country, winding up with a steep ascent of near a quarter of a mile, and priding himself on completing the distance from start to finish without stopping or even slackening his pace, then a cold bath, followed by breakfast''. He goes on to say that the reverend gentleman often wrote for at least twelve hours a day — all this presumably in addition to his clerical duties! Similarly Edward Newman, the editor of ''The Field'' for many years, usually commenced his day at three o'clock in the morning.

The energy and ability not only to undertake but to continue over a long period an enormous workload was not confined to the popular ''mass production'' writers, as is clear from the work of Richard Bowdler Sharpe. Whilst Newton was, academically speaking, a professional, Bowdler Sharpe was a professional in the modern sense of the word, that is his whole career was devoted entirely to ornithology. Like so many of his contemporaries he typified the Victorian work ethos: almost completely self-taught he held no university degrees, yet during his life he acquired an encyclopedic knowledge of birds on a world wide scale and rose to become the head of the bird section of the British Museum (Nat. Hist.).

He was born in London in 1847 but spent much of his early childhood in Brighton before going to Peterborough Grammar School where he won a Kings Scholarship. He was originally destined by his family for an army career but was quite unable to master mathematics to the standard required. In spite of his father's wish that he should try to gain a place at Oxford, Richard Bowdler Sharpe chose to go out and earn his living as soon as he was able. He had been an avid book collector since childhood and this maybe decided him to take a job with W.H. Smith, the booksellers, from there he moved to the antiquarian book seller, Bernard Quaritch, where he formed a lifelong friendship with Mr. Quaritch.

From an early age Richard Bowdler Sharpe had decided that he was going to make a career for himself in ornithology, a livelihood which would have been impossible a few years earlier. The Natural History Museum had only established itself as a separate entity from the British Museum in 1881, and the bird section within it had been organised in a very haphazard manner. The real beginning of Bowdler Sharpe's ornithological career was when the Zoological Society needed a librarian owing to its rapid expansion. Osbert Salvin and Philip Sclater recommended Bowdler Sharpe for this position and he was duly appointed. In 1864 he had already begun work for the first of his numerous books. "A monograph of the Kingfishers". This was published in fifteen parts between 1868 and 1871 with 121 coloured plates by John Gerrard Keulemans. This was an extremely beautiful book and was to set the pattern for a great many books which were published in Britain during the last quarter of the 19th and the first decade of the 20th century. These were generally in folio format with hand coloured plates and were frequently devoted to a single family of birds; usually colourful ones such as bee-eaters, sunbirds or rollers. In many cases the text was fairly scanty, especially concerning the life history and behaviour of the birds. The explanation for this lack of detailed information was that frequently the author had not seen the bird in the field. He often only had a skin from which to work and simply had to depend on whatever sketchy notes the collector had sent with the specimen.

In May 1872 George Grey, the Keeper of the bird section of the Museum died, and Bowdler Sharpe was appointed to this position. During this period vast tracts of the globe, hitherto completely unknown to ornithologists, were being scientifically explored by professional collectors. Many of the soldiers and administrators of the by now far flung British Empire were interested in natural history and huge numbers of skins and eggs were sent back home to England for examination and classification. There were also the great private collections of men such as Seebohm, Dresser and Hume. The numbers of specimens involved in these were huge, Octavius Hume's collection alone numbered some 82,000 skins. Nearly all these collections both from home and abroad were eventually offered to a museum, and such was Bowdler Sharpe's popularity and reputation that the British Museum was usually offered the best.

There were also at this time a number of special expeditions being organised by various scientific bodies to explore specific regions of interest. One of these was the Yarkand Mission, the bird collections of which were examined and written up by Richard Bowdler Sharpe himself.

Of all the tasks which he undertook, the greatest in scope and the most important was his contribution to "The Catalogue of Birds in the British Museum". This monumental work consisting of 27 octavo volumes was virtually a treatise on the birds of the whole world. It was published between 1874 and 1898 contained 387 coloured plates and is a unique work. Bowdler Sharpe wrote no less than eleven volumes himself besides being co-author of another two. This onerous task was made even more formidable since because of the pressure of work at the Museum he was forced to do nearly all his writing in his spare time.

In addition to his work on "The catalogue of Birds", he wrote also a great many papers and articles. Yet another major scientific work for the museum was "The Handlist of the Genera and Species of Birds", published in five volumes between 1899 and 1909. This was a long, complicated and difficult task which he completed only shortly before his death.

Amongst his "popular" books were "Sketchbook of British Birds" (1898) and "A Handbook to the Birds of Great Britain", four volumes, 1896-97.

"Monograph of the Paradiseidae, or Birds of Paradise and Ptilonorhynchidae or Bower Birds" was a magnificent work published in two volumes between 1891 and 1898. This contained 79 coloured plates by Gould and Hart, some of which were from Gould's "Birds of New Guinea" and some were produced especially for this work. With Claude Wyatt he was joint author of "A Monograph of the Hirundinidae or Family of Swallows" — two volumes 1885-1895 — containing 103 coloured plates, (although many copies exist with a number of the plates uncoloured).

In addition to the books of which he was author or co-author he undertook the work of finishing no fewer than four of John Gould's books and, following the death of Henry Seebohm, completed "A Monograph of the Thrushes". Seebohm in fact had hardly begun this book and although his name does not appear on the title page, Bowdler Sharpe was almost the sole author. He was editor of "Allen's Naturalist Library", and also wrote the birds section in "The History of the Collections contained in the Natural History Departments of the British Museum" (1904). He gained great personal satisfaction from writing this last contribution which was particularly well written; although again being undertaken mainly in his spare time it took him nearly two years to complete.

Besides his prodigious literary output he was the founder of the British Ornithologists Club, at whose dinners he was a regular and enthusiastic attender. He delivered numerous lectures, and was a keen amateur archeologist spending some of his holidays helping Lord Bolton's team with their excavations at Basing Castle. He enjoyed fishing and was a fair performer at tennis and cricket.

His death from pneumonia on Christmas Day in 1909 left a huge gap in British ornithological circles. He had become something of an institution and there were few facets of ornithology on which he had not left his mark. Those who knew him all remarked on his great learning, friendliness and good humour. His normal mode of address to one and all being "My dear boy...." with the single exception of Henry Dresser to whom in later years he could hardly bring himself to be civil.

He had a great propensity for practical jokes and the result of one can be seen to this day. When Eugene Oates's was working in the Museum and writing about the calls of the pied wagtail he happened to leave the manuscript open on the desk while he went off to lunch. Bowdler Sharpe and Ogilvie Grant passing by, noticed this, and Bowdler Sharpe said "Let's add something funny to Oates's description" and did so, fully expecting that it would be noticed when Oates came back from lunch. However it was not and in Volume 2 (Birds) on page 290 of "The Fauna of British India" there appears the statement: — "The note of this species is a penetrating POOH"!

At a time when professional ornithologists were still a very rare occurrence Lord Rothschild brought from Germany to his museum at Tring a remarkable man called Ernst Johannes Otto Hartert. Although the bulk of his writing was published in periodicals such as "Novitates Zoologicae" and "The Ibis" he was of great importance to British ornithology. His great work "Die Vogel der Palaarktischen Fauna" although written in German was largely the result of work done in England and was one of the most important taxonomical works yet to be published.

Ernst Hartert was born in Hamburg in 1859, but spent much of his childhood in East Prussia where he later made a special study of the birds of the Masurian Lakes and the Kurischehaff, one of the finest areas in Europe from which to watch migrating birds. Many of his observations made at this time were published in "The Ibis" some years later. His ornithological activities were not confined to Europe and he travelled far and wide to Nigeria, Sierra Leone, Penang, Sumatra, Assam and the Himalayas.

On his return to Germany he was commissioned to catalogue the famous Senckenberg Collection in Frankfurt. So well did he accomplish this assignment that he attracted the attention of Richard Bowdler Sharpe who invited him to write the section on the swifts and nightjars for "The Catalogue of Birds in the British Museum". After taking part in a collecting trip to Puerto Rico and Venezuela he returned to England in 1892 to take up the appointment of Director of Lord Rothschild's museum at Tring. Under his direction this wonderful collection, entirely financed and directed by Lord Rothschild, became recognised as one of the finest in Europe. For 38 years, until his retirement in 1930, Hartert built up and recorded this wonderful collection, his orderly painstaking methods contributing in no small way to the final result.

One highly important facet of Hartert's work at Tring lay in the fact that he firmly established the use of the trinominal system and the recognition of geographical forms of birds. His progressive attitude met with fierce opposition from the old school of taxonomists such as Newton, Dresser and Seebohm, who considered that many species which Hartert designated as geographical races or subspecies were in fact full species: an argument which led to a great deal of acrimony and confusion. What was required was a complete and fundamental revision of the whole nomenclatural system, and this was ultimately provided by Hartert.

He was an interesting character, pedantic with an obsessive regard for detail, he never lost his thick German accent throughout the many years he lived in England. In his appearance too he was typically Prussian, with an egg shaped head covered with short bristly tan-coloured hair. The first world war was a terrible time of tragedy for Hartert, his only son was killed fighting in the British army, whilst many of his relatives and closest friends were in Germany. After his retirement in 1930 he returned to Berlin and was allocated a room in the Museum where he continued to work until his death in 1933.

The relationship between himself and Lord Rothschild always remained formal: even after a close association of more than forty years Lord Rothschild always called him Doctor Hartert and Hartert invariably referred to his employer as Mr. or Lord Rothschild. Their occasional quarrels were legendary. Lord Rothschild would sometimes buy large and expensive mammalian exhibits when Hartert felt that such funds would be better used to purchase more interesting bird specimens. The resulting argument, usually carried out in German and behind closed doors, could, nevertheless, be clearly heard all over the Museum.

The great majority of 19th century ornithologists were either academics or men possessed of considerable private means. The fulltime professional was truly a "rara avis". One such was James Edmund Harting. Born in Chelsea in 1841 he was educated at Downside and London University. He had originally intended to become a solicitor but did not enjoy this profession and instead in 1871 became the natural history editor of "The Field", and by 1920 had contributed 2,326 articles to that journal. From 1871 to 1896 he combined his appointment with "The Field" with that of the editorship of "The Zoologist". Following its move to the new building in South Kensington in 1883, he was deputed to organise and build up the magnificent library at the British Museum (Natural History), and remained there until 1888. Subsequently he worked for the Linnean Society as librarian and assistant secretary until his retirement in 1902.

An enthusiastic falconer, John Harting founded the "New Hawking Club", an organisation for those who were interested in the sport but lacked the prestige or funds to be members of the old established hawking clubs. This interest led him to write the "Habits and Management of Hawks" and "Bibliotheca Accipitraria". The latter being an excellent catalogue of books on falconry. On ornithology, as distinct from falconry, he was the author of "The Birds of Middlesex" (1866), "Ornithology of Shakespeare" (1871), "A Handbook of British Birds" (1872), "Our Summer Migrants" (1878), and "The Birds of Hampstead" (1889). All of these were fairly run-of-the-mill works and contained illustrations of no great artistic merit.

In the matter of nomenclature and taxonomy James Harting was very much a conservative, an attitude which led to his resignation from the B.O.U. in 1918 because he felt that he could not longer go along with modern trends. He expressed this feeling in a letter published in "The Ibis" in 1912 in which he argued that dunlin should be spelt as "dunling" to be historically correct. Increasing deafness in later years greatly limited his activities and he died in January 1928.

VI

Establishing US Ornithology

Just as the development of ornithology in Britain in the 19th century was leading to the creation of the professional ornithologist, so in the United States much the same process was taking place. Amongst the most important of these professionals was Spencer Fullerton Baird, a man who was in many ways the counterpart of Richard Bowdler Sharpe. Like Bowdler Sharpe he became the secretary of the most important natural history museum in his country, namely the Smithsonian Institution. Through his great organisational ability, his capacity for cooperation with others and, above all, his helpfulness and encouragement to all actual or potential collectors, Baird succeded in forming at the Smithsonian a unique and marvellous collection. This called for unremitting effort on his part and he often worked for fifteen hours a day. During his twenty-eight years as Assistant Secretary and then Secretary at the Smithsonian the number of specimens in the collection rose from some 6,000 specimens to several hundred thousands, with a further 400,000 being distributed to other museums. At the time of his death in 1887 the Smithsonian contained no fewer than 2,500,000 specimens!

Spencer Fullerton Baird was born at Reading, Pennsylvania in February 1823. He attended Dickinson College graduating from there in 1840 at the very early age of seventeen, and then spent several years in New York studying natural history and medicine, although he never completed the medical course. In 1845 he returned to Dickinson College as Professor of Natural History, and in 1850 accepted the position of Assistant Secretary at the Smithsonian Institution. Throughout his time there he corresponded with anyone in the scientific community who might be able to assist the Institution in any way, amongst these were men such as J.J. Audubon, Thomas Brewer, William Gambel and John Cassin. He was always willing to send instructions for collecting specimens to amateurs and professionals alike as well as to military personnel stationed at outposts on the frontiers. His contacts with naturalists were not confined to his own countrymen, and he exchanged specimens, books and ideas with many others, including Richard Bowdler Sharpe in London.

The frontiers of the United States were steadily expanding throughout the 19th century, during this time the U.S. government sent out numerous expeditions to determine boundaries and make surveys concerning the practicability of building railway routes. Baird always tried to have a naturalist attached to these expeditions so that any new ornithological discovery could be properly observed and reported back to the Institution. His work-load was phenomenal. In 1860 alone, without the services of a secretary, he wrote no less than 3,050 letters to collectors and friends; answering enquiries from the public at large and to suppliers of equipment for expeditions in the field. He made a particular point of seeing that his collectors received the credit for their discoveries and that due acknowledgement should be made in the Smithsonian reports. In the case of some of the military personnel he even managed occasionally to obtain promotion for them, or like the Hungarian Xantus, rewarded by having a bird named after them (Xantus Murrelet).

Besides his duties at the Smithsonian he was appointed by Congress to be the first Commissioner of the United States Commission of Fish and Fisheries in 1871 and was also founder of the famous scientific station at Woods Hole, Massachusetts. His contribution to ornithological literature was not as large as that of some of his contemporaries but without his help and advice many books by others would not have been written.

Together with John Cassin and George Lawrence he was the author of "The Birds of North America", published in 1860 which was mainly based on Volume 9 of the U.S. Pacific

Color in true 1874 edition is by hand.

Railroad Survey. The single major book by which he is remembered is "A History of North American Birds" published in 1874 in three volumes with woodcuts and 64 plates either in black and white or chromo-lithographs. The co-authors were Thomas Brewer, Robert Ridgway, Elliott Coues and Theodore Gill.

By the time of his death in 1887 Spencer Fullerton Baird had succeeded in advancing the ornithological establishment in the United States from comparative insignificance to being widely recognised as one of the greatest in the world.

At the same time as Baird was working in Washington there was also, in the same city, another important naturalist, Elliott Coues. Along with a colourful personality, which as he grew older verged on the eccentric, Coues, like Baird, possessed a great capacity for work and was an active and enthusiastic collector and field observer. Born at Portsmouth in New Hampshire in September 1842, his family moved to Washington in 1854 where he received most of his education. In 1860 he was sent by Baird on a voyage to Labrador on the schooner "Charmer". One of the objects of this trip was to search for the great auk and, although in other respects this expedition was successful, Coues had to report that this bird had now clearly become extinct in North America — as indeed it was elsewhere. Despite his youth he displayed considerable independence of mind and showed no hesitation in differing from the opinions of Audubon concerning the Esquimo curlew. Unfortunately the journals which he kept throughout this voyage have been lost.

In 1861 he produced his first scientific paper, "A Monograph of the Tringae of North America", which displayed the same high standard of work which he maintained throughout his career. In this paper he named Baird's sandpiper from a collection by Robert Kennicott and Bernard Ross made at Great Slave Lake (Canada). In the same year he graduated from Columbia College (now the George Washington University) and, probably on the recommendation of Baird, transferred to the medical faculty of the same University, graduating from the medical school in 1863.

In 1864 he was commissioned as Assistant Surgeon, equivalent to a first lieutenant, in the United States army. His army service was spent largely in the South West, Kansas, New Mexico and Arizona, and here Coues started work on a book describing the birds of the area; but when the manuscript was almost ready for publication he suddenly destroyed it, it is thought because Baird had criticised it. Coues spent seventeen years in the army during which time he collected a great deal of material which he sent back to the east. Acting on Baird's instructions, he also collected as many bird species as possible between the Rio Grande and Rio Colorado, whilst carrying out his ordinary military duties. During such expeditions he was frequently mounted on his favourite biscuit-coloured mule which he named 'Jenny Lind' because, as he said, "She had such a musical bray". He would ride away from the column, by himself, and generally return with his saddlebags filled with specimens which he carefully prepared in camp each night.

His great work "The Key to North American Birds", written while he was stationed in Maryland, was first published in 1872. This book can truly be described as a real milestone in the history of ornithological publication. It was a single volume encyclopedia of ornithological practice and the birds of the U.S.A. Commencing with a history of ornithology in North America — up to that date — there followed a chapter of instruction for collecting and preserving birds and collecting and preparing eggs. A further section was devoted to classification, and another on avian anatomy written and described in considerable detail. The final section consisted of a species account illustrated by six black and white steel engravings and some 250 woodcuts.

This ran to five editions and the anatomical and historical sections were added to considerably after the first edition. There was also a separate publication entitled "Field Ornithology" which expanded the sections on classifications and collecting etc. His literary output was enormous but as yet no complete bibliography of his work has been compiled. What Coues hoped would be his greatest work was his unfinished "Universal Bibliography of Ornithology", in which he intended to list all known ornithological titles. In his later years he seemed to be almost relieved that this mammoth task had been beyond him for he wrote: "It takes a sort of inspired idiot to be a good bibliographer, and his inspiration is as dangerous a gift as the appetite of the gambler or dipsomaniac.... perhaps it was lucky for me that I was forcibly divorced from my bibliographical mania: at any rate, the years have cured me of that habit, and I shall never be spellbound in that way".

In 1873 to 1874 he was the naturalist and secretary of the Border Commission which determined the boundary between the U.S.A. and Canada. This resulted in another of his major works, "The Birds of the Northwest" published in 1874, following in 1878 by "Birds of the Colorado Valley", this contained the first four parts of his bibliography referred to above. A self-confident and often irascible man, Coues was not always an easy character to work with. In "Birds of the Colorado Valley" he was harshly dismissive of those who did not agree with his theories on hibernation. His overlong, and sometimes irrelevant, footnotes in some of his books irritate many of his readers and another, less innocent fault, was his habit of making handwritten notes on priceless manuscripts. He also on occasions altered quotations from other peoples' work without appraising the reader of the fact that the meaning of the quotation was not precisely as the original author intended. Certainly he later revealed a streak of ruthlessness from his early days, when he admitted to the practice of body-snatching whilst a medical student.

Coues resigned his commission in 1881 to become the full-time Secretary of the Geological and Geographical Survey of the Territories, in addition he was also appointed Professor of Anatomy at the George Washington University. In 1884 he visited London, at the invitation of Richard Bowdler Sharpe, and gave a talk on the trinominal system of which he was a pioneer and leading proponent. The meeting was well attended and a lively discussion ensued with a lot of vocal opposition. His advocacy of the trinominal system was described by Stresemann as follows: "In 1884 a spark flew over the ocean and kindled a fire in Europe, immediately a general alarm was sounded to preserve the venerable structure of Linnean nomenclature and, if possible, to smother the conflagration entirely". Although Coues was one of the most capable exponents of the trinominal system he was not its originator, it was put forward by Sundevall (a Swede) and Schlegel (a German).

Elliott Coues played an important part in the compiling of the first "American Ornithologists Union" check-list of the birds of the United States. In this the trinominal system was used, but he was extremely angry when at a later date the A.O.U. rejected some of his classifications. As he grew older he tended to indulge in private vendettas against other ornithologists. Robert Ridgway was one of his particular "bete noires", and apparently out of sheer envy he accused Ridgway openly of plagiarism; the situation was exacerbated when Baird proposed Ridgway as Curator of Birds at the U.S. National Museum. "It must be well known to you that our friend Ridgway never yet went twice over any group of birds with the same result", he wrote spitefully to Joel A. Allen in the October of 1885, "and those he studied longest are the most like a game of hop-scotch to him". Probably his most notable feud, to the spectator at any rate, was the "Sparrow War" which got under way in 1877. Coues bitterly opposed the practice of releasing European sparrows into the U.S.A. on the grounds that they would have a harmful effect on the indigenous birds. On the other hand, Thomas Brewer, a Boston physician, journalist and ornithologist thought the introduction of the sparrows was an excellent idea. Bitter articles, notes to journals etc. were fired off by one side or the other, and although they had once been quite friendly the atmosphere between them became completely poisoned. Brewer declared that for Coues "A lie well stuck to being as good as the truth", and when Brewer died in 1880 Coues writing again to Joel A. Allen said, "We both know that Brewer was a cantankerous old ass at the time he had the good taste to fall asleep in Jesus". The argument continued to rage on for another few years by which time Coues had become embroiled in arguments and quarrels with two other ornithologists, Clinton Hart Merriam and Leonard Stejneger. He was however capable of great kindness as well going to great lengths to help the young artist Louis Agassiz Fuertes, acting as his mentor and helping him to get his paintings exhibited at an A.O.U. Congress and recommending his work to other authors.

In his later years Coues became greatly interested in the occult and theosophy, but lost most of his enthusiasm after a number of strong disagreements with various members of the sect. Apart from his ornithological work, in the last decade of his life he edited the journals of some of the great American explorers such as Lewis and Clarke, Zebulon M. Pike, the Henry and Thomson Journals and, finally, the diary of Francisco Garces. It was while searching for further Spanish manuscripts in Arizona and New Mexico in 1899 that his health began to fail, and he died of cancer soon after his return to Washington. With all his faults he was a very great naturalist and historian as well as doctor and scientist.

Although he was primarily a museum worker rather than an explorer or collector mention should be made of John Cassin. He was born in 1813, a Pennsylvania Quaker, he became a business man in Philadelphia, a city whose Academy of Natural Sciences had the finest collection of bird specimens in the U.S.A. at that time, (some 26,000 skins in all). Despite his business commitments he devoted a great deal of his time to ornithology of which he acquired over the years a very wide knowledge. Coues said of him that he was the only American ornithologist of his day who knew the avifauna of the old world as well as he knew that of the new. He made one important overseas expedition to Japan with Admiral Perry on his historic voyage of discovery in 1852-1854. Cassin's most important literary work was the publication of "Illustrations of the Birds of California, Texas, Oregon, British and Russian America". This was the first major work on the birds of the American west besides being a pioneering work in that trinomial names were used for the first time in the U S A. The book was published in ten parts between 1853 and 1856, and was illustrated with fifty chromolithograph plates by G.G. White and W.E. Hitchcock, though of rather indifferent quality. A preliminary part of a publication of the same title was issued in 1852 but was replaced by the work referred to above.

Apart from his books John Cassin contributed a great many papers to scientific journals, and also wrote up the reports of several expeditions who had sent back large quantities of scientific material. He died in 1869.

Amongst the number of outstanding ornithologists who significantly advanced the study of birds in North America during the last half of the 19th century was Daniel Giraud Elliot. He was born in New York City in 1835 and as a child exhibited a great interest in natural history and began to collect specimens of birds at an early age. It was intended by his family that he should enter Columbia College, but his health was considered to be too delicate for a rigorous academic career, and instead he embarked on an extensive series of travels, taken in the first instance for his health and later for research.

The earliest of his journeys was to the southern states of the U.S.A. and the West Indies, and in 1857 he visited Brazil. Later he visited Malta, Sicily, Cairo and the Middle East. Later, as the representative of the new American Museum he toured the museums of Paris and London making valuable contacts which were to be of great use to him in his work. In 1853 he returned to New York and began a lengthy series of papers on the birds in the Museum's collections. In 1888 he embarked on another expedition to Montana in search of bison. Not one living specimen was spotted throughout the entire journey, a terrible indication of how disastrously their numbers had declined since the opening up of the West.

In 1894 a new, and in many ways the most fruitful, phase of

his career commenced when he was appointed to be the curator of the new Field Museum in Chicago. It was on the Museum's behalf that he led a collecting trip to Somaliland in 1896, during which he contracted a serious illness. He resigned from the Field Museum soon afterwards, although he continued to maintain the closest contacts with that institution.

Elliot's first major ornithological work was "A Monograph of the Tetraoninae", published in five parts between 1864 and 1865, with 27 coloured plates by J. Wolf, W.S. Morgan and Elliot himself, although it is recorded that he was dissatisfied with his own artistic work. This was followed by "A Monograph of the Pittidae" published in 1867 with 31 coloured plates. A second, and enlarged, edition of this work was published in 1893-1895 with 51 coloured plates most of them done especially for this edition. Elliot went on to publish a series of fine monographs, including "A Monograph of the Paradiseidae" with 36 coloured plates in 1873: and "A Monograph of the Bucerotidae" with 57 coloured plates. These beautiful books were very much in the pattern of those published in Britain by H.E. Dresser and Seebohm. Elliott also wrote some excellent books of a more popular nature, "North American Shore Birds" (1895), "The Gallinaceous Game Birds of North America" (1897), and "The Wild Fowl of the United States and British Possessions" (1898): besides publishing many papers in such scientific journals as "The Ibis" and "Proceedings of the Zoological Society".

Following his return to New York he did little more ornithological work, devoting his whole attention to the study of mammals. He died in December 1915.

Just as during the 19th century the United States changed from a vast undeveloped and largely unexplored land into a dynamic industrial and commercial nation, so under the guidance of men such as Baird, Coues, Cassin and Elliot the burgeoning interest in natural history was fostered and expanded, and scientific ornithology firmly established upon a sound foundation.

Within a period of some fifty years the American

Ornithological Union was established, as was the American Museum of Natural History and the Smithsonian Institution; numerous bird clubs were created and flourished and a vast body of ornitholigical literature produced.

VII

The Collectors' Contribution

The latter part of the 19th century saw the emergence of the professional ornithologist, although such men, apart from the professional collectors who were in a somewhat different category, were only to be numbered in single figures. Alongside them was a large body of capable and energetic amateur ornithologists representing a wide range of activities in the ornithological field. Usually they were men of independent means who were willing to devote most of their spare time and not a little of their money in the pursuit of their great interest in, not to say obsession with, birds. The tangible evidence of the work of many of these enthusiasts remains today in a whole series of books, often beautifully illustrated with hand coloured plates, and many of them published privately at considerable cost to their authors.

Typical of this class of dedicated amateur was Henry Eeles Dresser. Born at Thirsk in Yorkshire in 1838 he was educated at Bromley in Kent before being sent in 1852 to school in Hamburg. His family were timber merchants and as much of their trade was with northern Europe he was then dispatched to Uppsala in Sweden in order to learn the language. On his way to Uppsala however he stopped at Gothenburg for a time where he learned the rudiments of the craft of taxidermy. In 1856 he visited St. Petersburg (Leningrad) and from thence went on to Finland. In 1858 whilst travelling around the Baltic he collected a waxwing's nest which he found at Oulu, the first oologist collector to do so.

In 1859 he visited Canada and in 1863, Texas. From 1864 to 1870 he travelled abroad every year visiting Spain, Turkey, Austria, Italy, Serbia, Roumania, Bulgaria and Russia. During the course of these journeys he gained an extensive knowledge of European birds as well as becoming fluent in several northern European languages, a facility which proved to be of the greatest use to him in his later ornithological activities.

His first, and greatest, book was "A History of the Birds of Europe" published in 84 parts between 1871 and 1881, and containing 633 hand-coloured plates. There was, in addition, a supplement consisting of nine parts with a further 89 plates published between 1895 and 1896. These plates were executed by J.G. Keulemans, Joseph Wolf and E. Neale. The first eight volumes cost £50.00, and the supplement nine guineas (£9.45): a considerable sum of money at that time but today they would fetch in excess of £7,000. Richard Bowdler Sharpe assisted Dresser in the preparation of the first twelve parts but a mutual antipathy seemed to have developed between the two men and this was to be their only collaboration. "A History of the Birds of Europe" was an altogether more complete and professional study than any previous work on the birds of a region which is now referred to as the Western Palearctic. A much less sumptuous book by Dresser, but of great ornithological value is "A Manual of Palearctic Birds", published in two parts in 1902 and 1903. This single volume contains within its 922 pages good descriptions of all the birds of this region but has no illustrations. Nevertheless until the advent of the fieldguides in the 1950s it was of inestimable value as a means of identification.

Henry Dresser then went on to publish privately two beautiful books on two particularly attractive and colourful families of birds, the bee-eaters and the rollers. The first, "A Monograph of the Meropidae, or the Family of the Bee-eaters" was published in five parts between 1884 and 1886 and contained 34 hand-coloured plates drawn by J.G. Keulemans. The second book, equally fine, was "A Monograph of the Coraciidae, or Family of the Rollers", published in 1893 contained 27 hand-coloured plates again done by Keulmans.

The discovery during this period of many new species of birds which could be illustrated for the first time served as a great stimulus towards the production of many of these marvellous books. The commercial viability of such publications was made possible by the existence of a large and growing number of prosperous individuals who were prepared to subscribe to such finely illustrated books which could be really described as the prototypes of the 20th century "coffee-table" books. Since nearly all these works were issued in parts, with in many cases a lengthy gap between the first and last issues, many of these subscribers failed to complete their sets. This has resulted in many incomplete sets being offered for sale, these are often broken up and the illustrations framed and sold individually as pictures, a highly economic proposition at today's high prices.

Besides his enthusiasm and seemingly boundless energy, Henry Eeles Dresser brought a single-minded intensity to any task on which he was engaged. Ultra-conservative in his views he would never accept the existence of sub-species, a belief he firmly adhered to until the end of his life. Besides building up a large collection of bird skins and other specimens he was also a great egg collector and like many of his generation spent quite a lot of time at salesrooms such as those of Messrs. Stevens where vast quantities of such things were sold.

Those who knew him personally described him as possessing "a striking appearance" with a sharp nose and intense features: in his later years he took to wearing a wig, and as his speech was forceful and rapid and generally accompanied by violent gestures, this often slipped badly. Another cause of amusement to his friends and acquaintances was his relationship with his wife. As she was rather vague and moved in spiritual circles and he was very precise and scientifically minded their conversations were apt to be on somewhat different wavelengths and a comment from one often received a reply on a totally different subject from the other.

At one stage of his life he spent much of his time at a house in Tenterden Street in the west end of London, this was rented for use by some prominent members of the British Ornithologists Union, and known to some as "The Ibideum".

Amongst others who had rooms there in his day were Lord Lilford, Osbert Salvin and Captain Shelley. Henry Dresser was a regular contributor to "The Ibis" and was secretary of the British Ornithologists Union from 1882 until 1888.

His great love of and interest in egg collecting resulted in his last major book, "Eggs of the Birds of Europe", published in 24 parts from 1905-1910, with 106 coloured photographic plates. This was the last of the great books on birds eggs to be published in Britain: the title-page states that it was "Published by the author (by special permission) at the Office of the Royal Society for the Protection of Birds". The huge collection of eggs on which he based this work, together with many bird skins was given to the Manchester Museum following his death in Cannes in November 1915.

Another gifted amateur who flourished during the latter part of the 19th century was Canon Henry Baker Tristram. He was a great character known, somewhat irreverently, to his friends as "The great gun of Durham" or "The Sacred Ibis". Born near Alnwick in Northumberland in 1822, the son of the local vicar, Henry Tristram was educated at Durham school and Lincoln College, Oxford from whence he graduated in 1844. He was ordained in 1845 and appointed curate of Morchard Bishop, but following a serious chest complaint he was advised by his doctors to live abroad and as a consequence spent two years in Bermuda as naval and military chaplain. In 1855 he visited Algeria and developed a special interest in the birds of North Africa and later of the Middle East. The Mediterranean climate, "bright, dry and invigorating" as he described it benefited his health enormously and he always maintained that his recovery was entirely due to these visits. He published a large number of papers in "The Ibis" on the birds of these regions and also a book describing his journeys entitled "Great Sahara" which was published in 1860 giving vivid descriptions of his travels therein. He made several trips to Israel, Jordan, Syria and Iraq which led eventually to the publication in 1884 of "Fauna and Flora of Palestine", his most important single work and the one by which he is best remembered.

His method of collecting birds, from which he earned his nickname, was on a truly gigantic scale. By 1889 he had amassed some 17,000 bird skins and rather than risk the dispersal of this valuable collection bequeathed it, in its entirety, to the Liverpool Museum. By the time the museum had finally catalogued the skins they numbered more than 20,000 items of some 6,000 differing species.

His considerable and important work in the field of North African and Middle Eastern ornithology was recognised by the naming of two birds after him, Tristram's Grackle and Tristram's Warbler. He actually discovered the grackle himself, and it was named after him by Philip Sclater (it not being considered ethical to name a bird after oneself). Tristram also identified and named the Cyprus warbler and the Marquesa's warbler.

Despite his early delicate health Canon Tristram, like many of his contemporaries, was remarkably tough; when at the age of 75 he had his leg broken by a kick from a vicious horse he was back in circulation within a few weeks, and remained in fair health until his death in March 1906.

Howard Saunder's reputation rests largely on a single book, but no account of notable Victorian amateur ornithologists can possibly ignore his work "An Illustrated Manual of British Birds" which followed directly on Yarrell's work and in fact retained most of his illustrations. It fulfilled admirably the need for a single volume reliable birdbook and was to be the standard work for many years. Published in 1889 with a second edition in 1899, during the last decade of the 19th century and the first two of the 20th, a large proportion of birdwatchers relied upon Howard Saunders' book in much the same way as their successors came to depend on the Witherby handbook.

He was born in London in September 1835 and was educated at Leatherhead and Rottingdean. He joined a firm of merchant bankers, and was sent by them to Peru where he worked in the port of Callao. Whilst in South America he devoted much of his spare time to archeological research as well as natural history.

He returned to Britain in 1862 but continued to go abroad regularly, especially to Spain about whose birds he became an acknowledged authority. He was also considered a world authority on the gulls and terns and wrote the section concerned with that family in "The Catalogue of Birds in the British Museum". An excellent and efficient committee man he became vice-president of the Zoological Society, vice-president of the Linnean Society besides, in 1901, becoming secretary of the British Ornithologists Union.

Howard Saunders was a popular and respected figure in ornithological circles; always generous with his time in helping others and in reading and correcting papers before publication. He died after a long illness in 1907.

One of the most outstanding and interesting personalities amongst the amateur naturalists and ornithologists at this time was Thomas Lyttleton Powys, later to become Lord Lilford. Born in London in 1833 his lifelong interest in the study of birds manifested itself at an early age. He was educated at Harrow where he soon gained a certain amount of notoriety by keeping two bitterns in his locker — quite a feat since they are largely fish-eating birds. Although he possessed a certain faculty for foreign languages, aided no doubt by a good ear for music, he was not a great scholar. On leaving Harrow he spent a while in Lausanne where he built up a collection of alpine birds. In 1852 he went up to Christchurch, Oxford, where he spent most of his time out in the countryside either shooting or hunting for birds. His unconventional habits soon brought him into conflict with the college authorities who did not welcome an undergraduate who kept badgers and snakes in his rooms, and he left the University without taking his degree.

Although he had the reputation of being something of a practical joker Lord Lilford, like all the true eccentrics, probably considered his lifestyle to be perfectly normal. Returning to London he took rooms in Down Street just off

Piccadilly, where he caused quite a furore one day when he arrived in a hansom cab with an eagle owl in a cage on the roof and two armadillos inside with him, one of which promptly escaped and ate one of his neighbour's cats. He now developed a great interest in aviculture, a fascination which remained with him for the rest of his life. He was especially interested in the birds of prey and kept a large number both for falconry and for pleasure at his home in Northamptonshire. Among his collection of birds perhaps the most impressive were several free-flying tame lammergeiers, which must have provided quite a spectacle as they soared around the countryside.

After leaving Oxford Lilford served for a while in the militia before embarking upon a series of journeys around the Mediterranean, visiting ports in Spain, Italy and Greece on his private yacht. He particularly loved Spain and visited it frequently: he spoke Spanish perfectly and was also fluent in some of the regional dialects such as Andaluz.

As a consequence of these travels Lord Lilford contributed a series of papers on Spanish ornithology to "The Ibis".

Following his succession to the title of Lord Lilford in 1861 more funds became available to him which enabled him to enlarge his already considerable collections. His passion for falconry grew to the point where he needed the services of two full time falconers. From early middle age he became increasingly afflicted by a form of hereditary gout which had apparently affected his forebears on both sides of the family. He was often in great pain and became increasingly immobile until he was finally completely confined to a wheel-chair. Sometimes there were days when he was unable even to hold a pen and his wife had to write his letters and do other secretarial work for him.

In 1885 he commenced what he referred to as his "great work" — "The Coloured Figures of the Birds of the British Islands". He saw, quite rightly, that there was a gap between such rare and expensive books as Gould's "Birds of Great Britain" and those of Yarrell and Howard Saunders which lacked any coloured plates at all. What was required was a book which contained good coloured illustrations at a reasonable price.

Joseph Wolf was Lord Lilford's first choice of illustrator for his book, but by now Wolf was about sixty-five and did not wish to commit himself to the task of painting over 400 plates, an undertaking which could easily take ten years or more. Instead he worked with J.G. Keulemans who completed some 125 plates before ill-health prevented him from continuing with the work. It became essential to find another artist, and this time Lord Lilford chose the comparatively unknown Archibald Thorburn to complete the bulk of the unfinished plates. In all Thorburn painted 268 plates, the few remaining ones being executed by George Lodge, E. Neale, J. Foster and J. Smit.

Throughout the production of this book Lord Lilford insisted on maintaining the strictest supervision over the painting of the plates, and although frequently free with his praise he was not afraid to criticise points which he considered to be inaccurate or not clearly illustrated. Besides his very great personal knowledge of birds he had the additional advantage that in many instances he had the live birds in his aviaries to which he could refer. In a letter to Archibald Thorburn dated 21st July 1891 he wrote: "The two guillemot pictures and the little egrets are quite perfect, but I find that the naked space between the eye and the beak in my egrets is tinged with a shade of green. Do what you like about any alteration, as the slatey tint without the green is no doubt correct in certain stages of this species. I believe that you took the original sketches from life here, and no doubt the birds may have changed slightly since you saw them". More critically on another occasion he wrote to Thorburn. "The wood sandpiper is the only one I can really find fault with, he ought to be on the ground, his legs are rather too long, and his whole coloration is too light".

Lord Lilford wrote the whole of the text for "The Coloured Figures" himself keeping it brief and to the point. In order to keep the price of the book as low as possible he contributed some £15,000 of his own money towards the cost of the publication. The cost to the purchaser was 9/6d (47p) per part, and 12/- (60p) per part for the second edition. There were 36 parts in each edition.

Although this is the book for which he is best remembered he also wrote many papers for "The Ibis" and "The Zoologist" besides "Notes on the Birds of Northamptonshire and Neighbourhood" published in two volumes in 1895.

Always generous towards ornithological enterprises, Lord Lilford contributed considerable sums of money to the British Ornithologists Union whenever funds were short for the publication of "The Ibis", personally paying for many of the coloured plates in the early numbers of which there was always at least one in each issue. Present readers of "The Ibis" may not be aware that until 1912 it was customary for each volume of this publication to contain between ten and seventeen coloured plates. Lord Lilford's generosity even extended to offering financial assistance to Henry Dresser for his "The Birds of Europe" when it appeared likely that it might result in a considerable monetary loss to its author — fortunately such assistance was not required. There are numerous examples of anonymous gifts attributable to him to all sorts of individuals and institutions for a great variety of causes.

His last years were spent in constant pain and he died, following a stoke, on 17th June 1896. His memorial "The Poloured Figures of the Birds of the British Islands" was published between 1885 and 1897, the last three parts being completed by Osbert Salvin. A second edition, also edited by Osbert Salvin, was begun in 1891 and completed at the same time as the first. This concurrent issue of parts of two editions has resulted in a great many sets being offered for sale which are amalgams of both editions, and to add to the bibliographer's problems, some of the plates were reissued after improvements. For anyone wishing to know more about this book, "Handbook to Lord Lilford's Coloured Figures of the Birds of the British Islands", by Hugh S. Gladstone, (London 1917) is a mine of detailed information, listing amongst other details quantities printed of each part and the dates of issue and artists of each plate.

Although "Coloured Figures of the Birds of the British

Islands'' was not Archibald Thorburn's first commission, it was the one which brought his name to the fore as an ornithological artist, largely due to Lord Lilford's great influence amongst the ornithologists of the day.

He was the son of Robert Thorburn, miniaturist to Queen Victoria, whose portrait he painted three times, as well as a favourite picture of the Prince Consort. Archibald Thorburn was born at Lasswade in Scotland in May 1860, and soon showed a particular talent for drawing and painting, especially in the medium of watercolours. When he was twenty-two his first published illustrations appeared in ''Sketches of Bird Life'' by J.E. Harting, and this was followed by a commission for 144 plates for Swaysland's ''Familiar Wild Birds''. In 1885 he moved down to London from Scotland where he met, and was much influenced by, Joseph Wolf, one of the finest natural history artists in the world at that time. In 1887 he commenced his work for Lord Lilford. As a result of spending much time sketching his subjects in the field his work was not only meticulously accurate but the birds appeared life-like as well, being carefully and closely observed and not merely taken from dead study skins.

Apart from his 4 volume work ''British Birds'', published between 1915 and 1918, and a smaller book bearing the same title and also in 4 volumes which appeared in 1925, his illustrations were published in many other books such as ''A Monograph of the Pheasants'' by C.W. Beebe published between 1918 and 1922.

That great artist and contemporary, George Lodge, wrote of him ''Thorburn was the greatest ornithological artist, superior to all contemporaries in the same field, with a wonderful gift of placing his bird subjects in harmonious surroundings''. Few have equalled and none surpassed his work. He died after a long illness in October 1935.

Another keen Victorian amateur ornithologist, and almost an exact contemporary of Lord Lilford, was Henry Seebohm. Born in Bradford in 1832 he, like Lord Lilford, early developed a fervid interest in natural history in general and birds in particular. He was a successful Sheffield steelmaster and as his financial resources increased he was able to apply considerable sums of money for expeditions to Greece, Asia Minor, Scandinavia, Germany and above all Russia.

It is Seebohm's journeys to Russia, and particulary to Siberia, which remain of most interest to the ornithologist today. The first expedition on which he was accompanied by John Harvie-Brown was made in 1875, their intention being to explore the route along the valley of the Petchora river in northern Russia. The main pupose of this trip was to find and collect the eggs of any bird species which at this time were unknown — at least in Britain: specimens and eggs of the grey plover, little stint and Petchora pipit were among those which the party collected. Their account of the birds they saw and collected gives a vivid picture not only of the countryside of Czarist Russia but also the typical attitudes, motives and methods of the Victorian ornithologist. To them the collecting of specimens of almost everything they saw was the primary object of all their fieldwork. It is easy to depreciate this as mere slaughter, much of it totally unnecessary, but it must not be overlooked that barely a century ago there was still much that was unknown concerning the existence and distribution of species and subspecies, and that expert photography and well illustrated handbooks, which we take for granted, simply did not exist. It should also be remembered that many of the marvellously comprehensive collections of birdskins in our museums, frequently referred to today, date back to those times.

There were, undeniably, a number of people at this time who did slaughter and collect every bird and egg in quantities which were quite indefensible on any grounds except those of greed, but Henry Seebohm was not one of these, his primary objective being to form a good working collection.

During the expedition to the Petchora he found the nest of Bewick's swan for the first time, and also collected 110 species, although many of these have subsequently been reclassified during this century. He was greatly excited to discover the Petchora pipit (Anthus gustavi) believing it to be a species entirely new to science. Henry Dresser confirmed this and named it Anthus seebohmii in his honour. However, much to Seebohm's disappointment, it had been discovered not long before and named by Swinhoe who had taken specimens on migration at Amoy in China.

Fired by the success of the Petchora expedition, Seebohm determined to travel even further eastwards into Asiatic Siberia and, accordingly, set off again in March 1877, this time with Captain Wiggins as his companion. They went first to St. Petersburg (Leningrad) in order to obtain the necessary permits and to make arrangements for transport and supplies, before setting off, via Moscow, to Nizhny Novgorod (Gorki). Here they left the railway and continued their journey by horse drawn sledges. Their route covered some 3,000 miles through the Urals and beyond but they saw disappointingly little in the way of birdlife. Travelling on through Omsk and Tomsk they arrived at Krasnoyarsk on 2nd April and eventually reached the Kureika river where their boat "The Thames" had been moored throughout the winter. In six weeks they had sledged some 3,240 miles, and Seebohm calculated that their journey had required the use of 1,100 horses, 40 reindeer and 18 dogs — at a total cost to them of £87 each!

Soon after their arrival at the Kureika river their boat grounded completely and they were unable to free her, this meant that they were unable to reach the mouth of the river and continue on up the Yenesei, the ultimate goal of the expedition and thus missed seeing many species of birds which they would otherwise have done. Nevertheless as Seebohm himself wrote later, the trip was, "on the whole satisfactory", and as a result of it he returned with a sizeable collection of both birds and their eggs. His two books on his Russian journeys, "Siberia in Europe" (1880) and "Siberia in Asia" (1882) recount the story of these expeditions: in 1901 they were combined into a single volume under the title of "The Birds of Siberia — a Record of a Naturalists Visits to the Valleys of Petchora and Yenesei".

Seebohm was also greatly interested in bird migration, and there are frequent references to this in his writings, with careful notes of the dates of arrival and departure etc. In furtherance of this interest he spent a while on the island of Heligoland where one of the first bird observatories in the world had been created by Heinrich Gatke. He also visited South Africa in order to watch the European migrants in their winter quarters, and to study the waders, a group of birds about which he was to write a monograph in 1888, "The Geographical Distribution of the Family Charadriidae, or the Plovers, Sandpipers, Snipes and their Allies". This was one of his greatest books although it is rather underestimated today. It contained 21 coloured plates by Keulemans, although Seebohm presented a number of copies of this work to his friends which were without these plates. There were two printings of which the second is distinctly the better.

An earlier work, now rather out of favour, was "A History of British Birds", published between 1883 — 1885, in six parts at £1 1/- (£1.05) each. This was primarily a book on eggs and not a treatise on British birds as its title might seem to imply. A more straightforward book on eggs by Seebohm was "Coloured Figures of Eggs of British Birds", published in one volume in 1896, a year after Seebohm's death, and Richard Bowdler Sharpe was responsible for much of the text.

Seebohm's most unsatisfactory book was "The Birds of the Japanese Empire" which was published in 1890. Whereas in his other books he had written with confidence of personal knowledge of the country and its birds this book was written without the benefit of either.

One of Seebohm's greatest ambitions was to find an original method of classification, but the system which he finally advocated proved to be quite unworkable in practice. The system was almost as cumbersome as the title of the book in which he attempted to explain its workings — "Classification of Birds, an Attempt to Classify the Subclasses, Orders and Suborders and some of the Families of Existing Birds". This was published in 1890 with a supplement in 1895 and the general appraisal of this book was as adverse as that of Professor Newton, whose caustic comments have already been mentioned. His final book was perhaps his best and certainly the most attractive, "A Monograph of the Turdidae, or Family of Thrushes" (issued in thirteen parts) in two volumes published in 1902. Seebohm planned this book and commenced writing the text but after his death in 1895 it was again completed by Richard Bowdler Sharpe — a factor which may have contributed to its success.

Nevertheless Henry Seebohm was an ornithologist of some considerable stature as well as possessing a kind and likeable character.

On rare occasions the British Ornithologists Union awards a gold medal "as a signal honour for distinguished ornithological work". This medal was first struck in 1922 to commemorate two of the Unions founder members, Frederick Du Cane Godman and Osbert Salvin. Since so much of their work was shared it seems only appropriate that these two talented amateur ornithologists should be considered together.

Frederick Du Cane Godman was born in January 1834 and was educated at Eton and Trinity College, Cambridge. However before actually going up to Cambridge he began his career in a somewhat spectacular manner. Accompanied by his tutor, he was sent by his parents on a Mediterranean tour which included Southern Spain, Greece and Istanbul. Arriving at Istambul his tutor suddenly announced his intention of returning to England, this he promptly did, leaving his eighteen year old pupil stranded with literally only a sovereign in his pocket. With the assistance of a friendly hotel proprietor Godman was able to obtain some funds and then made the acquaintance of the British Consul in Trebizond, with whom he embarked on an extensive cruise around the Black Sea. When visiting the site of Sestos, Godman, a somewhat impulsive character suddenly realised that this was the very spot from which Leander was supposed to have swum the Hellespont, so he instantly dived in and swam across to Asia himself. His journey home was equally adventurous, he was befriended by two soldiers and with them rode through Greece to Corinth from where he took a boat to Trieste and returned home via Vienna and Dresden. Thus by the time he arrived at the University in 1853 he had already seen more of the world than most of his contemporaries were ever likely to do.

He had always taken a keen interest in natural history and soon met others at the University who were similarly likeminded, chief amongst them being Osbert Salvin, with whom he formed a lifelong friendship, as well as Alfred and Edward Newton. In pursuance of their hobby Godman and Salvin spent much of their time with a Mr. Baker, who had a taxidermist shop in the Trumpington Road, and under whose direction they practised and improved their taxidermic skills. Godman was not solely interested in birds and spent many of his days in the fens and the surrounding countryside collecting butterflies and moths.

In 1855 he went on a trip to Italy and whilst there, seemingly quite unaffected by the cataclysmic events which had occurred during his previous visit to Asia Minor, decided to proceed, quite on his own, to the Crimea. He arrived in time to watch, from a neighbouring hillside, the capture of Sevastapol, in which his brother who was a captain in the 5th Dragoons took part. He was by no means the only spectator on this occasion, there are numerous records in the journals and diaries of the period of men occasionally accompanied by their wives, who came out from England during the Crimean War in order to see the armies in action.

Godman's first expedition made with the express purpose of collecting birds and eggs was in 1857 to northern Norway in the company of another brother, Percy, who was also an enthusiastic ornithologist. After some weeks in Norway they made their way to Muonioniska in Finland where they were joined by John Wolley. They then continued onwards to Stockholm and St. Petersburg (Leningrad), Moscow and Nizhny Novgorod (Gorki).

In 1861 he and Salvin set off for Jamaica and together went

on to visit Belize and Guatemala; this journey marked the start of their long connection with central America which was to culminate in the publication of their "Biologia Centrali-Americana". This was a joint venture on the natural history of central America which was finally published between 1879 and 1904 running in all to 257 parts. It was an enormous undertaking and uniquely for such a huge work they published it privately. The four volumes which were devoted to birds together with those on the diurnal lepidoptera, they wrote entirely themselves, but other naturalists who were specialists in their particular fields were invited to contribute to the sections on plants, reptiles and molluscs etc. This was, in every sense, a great work, and its 52 volumes serve today as a permanent memorial to the two men who were its architects and authors.

As he grew older Godman spent more of his time on entomology leaving much of the ornithological work to Salvin. He continued to travel widely, making collecting trips to the Azores, Madeira, the Canary Isles, India, Ceylon, South Africa as well as much of Europe. His last great ornithological publication, and one which he had long contemplated doing, was "A Monograph of the Petrels (order Tubinares) ", published between 1907 and 1910 in 2 volumes with 105 coloured plates by J.G. Keulemans. By this stage of his life so great was the pressure of his other work that he was finally forced to obtain the services of Richard Bowdler Sharpe in order to get it completed.

Besides his absence on numerous collecting trips, as well as his considerable output of books and articles, throughout his life he devoted much of his time to the British Ornithologists Union, of which he was secretary from 1870 until 1882 and again from 1889 to 1897, and president from 1897 until 1913.

A man of broad interests he was a great fisherman, hunted regularly and was an accomplished deer stalker, on one occasion stalking and shooting eight stags within a single day. Intellectually he was something of a polymath since in addition to his naturalist activities he was an authority on horticultural matters, a knowledgeable and discerning collector of ceramics, besides being a world authority on early Persian ware. Although his name is invariably connected with that of Osbert Salvin, Godman's personal contribution to ornithology should not be overlooked. He died in February 1919.

His friend and colleague, Osbert Salvin, was born near Haslemere in 1835, and was educated at Winchester and Trinity. His first ornithological expedition, in 1857, was to Algeria and Tunisia, in the company of Canon Tristram and Wilfred Hudleston. Like Godman he was one of the founder members of the British Ornithologists Union, and was the joint author (with P.L. Sclater) of the first paper in the first issue of "The Ibis" in 1859. This was entitled "On the Ornithology of Central America", a geographical region of particular interest to them since both men were to concentrate their studies on neotropical birds, i.e. the birds of south and central America.

In addition to his joint editorship of the "Biologia Centrali-Americana", Salvin was the editor of "The Ibis" from 1871 until 1882, and the treasurer of the British Ornithologists Union in 1897-1898. He was appointed to be the curator of the Strickland Bequest in the University of Cambridge, and compiled a catalogue of that great collection which was published by the Cambridge University Press in 1882.

Although he was primarily interested in the birds of central and south America, he had, nevertheless, a very wide general ornithological knowledge, particularly concerning the humming-birds, the petrels and shearwaters: and his expert knowledge of these families of birds led to his being invited to write the sections relating to them in the great "Catalogue of Birds in the British Museum". Following the death of his friend, Lord Lilford, Salvin undertook to complete his "Coloured Figures of the Birds of the British Islands". This was to be his last major ornithological work and he died in June 1898 at the family home in Sussex.

Even allowing for the golden haze which has now settled over the Victorian era, it could be said, on the whole, to have been a fairly peaceful one. Unfortunately for the wildlife of that time, sport began to take the place of war, and much of the sport began to take the place of war, and much of the aggression that might have been expended upon a human adversary was diverted into bloodsports. Pre-eminent amongst these pastimes was shooting; to be considered a "good shot" was a highly regarded social accomplishment. Royalty led the way, from the Prince of Wales on his Sandringham estates downwards enormous battues were the order of the day, and the walls of the stately homes were bedecked with the products of this slaughter. In imitation of their "betters", most Victorian house-holds boasted a case of stuffed birds and mammals — it was truly the golden age of the taxidermist! This fashion for possessing stuffed birds, often of dubious provenance, assumed for a few collectors, such as Edward Thomas Booth, positively manic proportions, and to satisfy this need tens of thousands of rare birds in the British Isles were shot by these collectors or their agents.

There were however some amongst their numbers who collected unmounted skins purely for scientific purposes, and of these private collectors probably none did more serious systematic work than Philip Lutley Sclater who, during the course of his lifetime, described some 1067 new species, 135 new genera, and two new families of American birds. In recognition of his highly important work, on mainly neo-tropical birds, Philip Sclater was accorded the singular honour of having a whole Bulletin (no. 49) of the United States National Museum (Smithsonian 1896) devoted entirely to a detailed bibliography of all his published writings up to that year, together with a biographical sketch — the only non-American ever to be so honoured.

He was born in Hampshire in 1829. He went to Winchester School and whilst still under the age of sixteen was elected a scholar of Corpus Christi College, Oxford, but because of his extreme youth was unable to go up to the University until 1846. Here he met Hugh Stickland, then Reader in Geology at Oxford, who took him under his wing and gave him a sound foundation in scientific ornithology. In 1849 Sclater gained a first class degree in mathematics and a pass in classics, these subjects and divinity being the only recognised disciplines at that time. After graduating he stayed on at Oxford for a further two years and took his M.A. in 1851. During this period he studied, amongst other things, modern languages becoming fluent in French, German and Italian. He managed to spend quite a lot of his time in Paris where he became friendly with Prince Charles Bonaparte who was one of the leading ornithologists of the day.

In 1851 Sclater entered himself for the bar, and eventually joined the Western circuit where he practised for several years. He continued to travel as much as possible and in 1856 visited the United States and Canada for the first time, traversing that continent by means of train, steamboat and canoe. In 1859 he went to Tunis and eastern Algeria and in April of that year was unanimously elected to suceed D.W. Mitchell as secretary of the Zoological Society, a position he held for 43 years. For a time he acted as private secretary to his brother who held office under Disraeli and was offered a permanent post in the Civil Service, but this he declined preferring instead to concentrate his efforts on the Zoological Society. Under his ministry the fortunes of that august body underwent a considerable improvement, debts were repaid and the membership almost doubled. He was a founder member of the British Ornithologists Union and the first editor of "The Ibis", a position he held for some forty (but not consecutive) years. He had a long and valuable connection with the Royal Society for the Advancement of Science and also with the Royal Geographical Society: and it was often largely through his influence with the latter Society that geographical expeditions were made to such little known places as Kerguelen and Rodriguez, and that such expeditions were generally accompanied by a competent naturalist.

Despite these activities he always retained his interest in neotropical birds, as was demonstrated by his participation, with Osbert Salvin, in the publication of "Exotic Ornithology" issued between 1866 and 1869 with 100 excellent hand coloured plates by J. Smit. This very beautiful work illustrated and described 104 rare or newly discovered species from South America. Another important work on the birds of that continent written by Sclater, this time in collaboration with W.H. Hudson, was "Argentine Ornithology", published in 1888-1889 in 2 volumes with 20 hand coloured plates by J.G. Keulemans. A further work, this time as sole author, was "A Monograph of the Jacamars and Puff-birds", published in 1879-1882 in seven parts and containing 55 hand coloured plates again by Keulemans.

Philip Lutley Sclater's literary output was, by any standards, extremely impressive, amounting in all to some 1,300 articles and books, all of them of a consistently high calibre. He was the author of a large proportion of the three volumes concerned with the woodpeckers and some of the families amongst the order passeriformes for the "Catalogue of Birds" in the British Museum. Another important work was "A Monograph of the Birds Forming the Tanagrine Genus Calliste", this was published in 1857-1858 and contained 46 hand coloured plates.

He retired in 1903 and went to live at his country home near Odiham in Hampshire, here he hunted regularly and followed other country pursuits of the day. In his latter years he became rather absent-minded and was described as possessing "pinched features and a scrubby beard like Don Quixote". Like many others, towards the end of his life he became very set in his ways, always a keen taxonomist, he was bitterly critical of the nomenclatural changes proposed in "The Handlist of British Birds" by Hartert, Jourdain, Ticehurst and Witherby, His objections to these changes would seem to have been based mainly on the fact that he felt it would take a long time to get used to and comply with the new international rules, rather than that they were incorrect per se. He died, following a carriage accident, in June 1913.

No account of the great amateurs in the field of ornithology would be complete without some mention of the work of Lord

John Ray "The Father of British Ornithology".

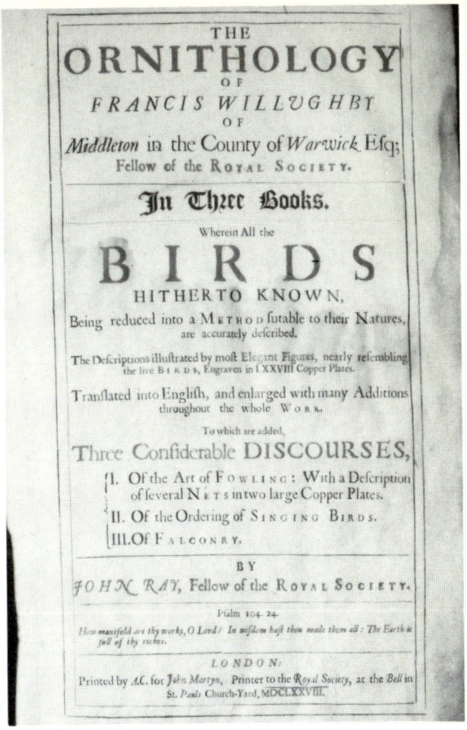

THE
ORNITHOLOGY
OF
FRANCIS WILLUGHBY
OF
Middleton in the County of *Warwick* Esq;
Fellow of the ROYAL SOCIETY.

In Three Books.

Wherein All the
BIRDS
HITHERTO KNOWN,
Being reduced into a METHOD sutable to their Natures, are accurately described.

The Descriptions illustrated by most Elegant Figures, nearly resembling the live BIRDS, Engraven in LXXVIII Copper Plates.

Translated into English, and enlarged with many Additions throughout the whole WORK.

To which are added,
Three Considerable DISCOURSES,
I. Of the Art of FOWLING: With a Description of several NETS in two large Copper Plates.
II. Of the Ordering of SINGING BIRDS.
III. Of FALCONRY.

BY
JOHN RAY, Fellow of the ROYAL SOCIETY.

Psalm 104. 24.
How manifold are thy works, O Lord! In wisdom hast thou made them all: The Earth is full of thy riches.

LONDON:
Printed by *A.C.* for *John Martyn*, Printer to the *Royal Society*, at the *Bell* in St. *Pauls* Church-Yard, MDCLXXVIII.

2. Willughby's Ornithology is one of the earliest ornithological handbooks. Note the "Three Considerable Discourses" by John Ray.

George Montagu, whose accurate and painstaking rk has never received the credit rightly due to it.

4. Montagu's "Ornithological Dictionary" of 1802 which clearly showed the progress made since Willughby.

5. Above Left. John James Audubon who illustrated and published one of the world's greatest bird books.

6. Above Right. Alexander Wilson, a pioneer of serious ornithology in the U.S.A. whose reputation was later eclipsed by Audubon.

7. A mid 13th century Latin manuscript, now in the Vatican Library, containing recognisable but innaccurate paintings of birds.

Above Left. Spencer Fullerton Baird,
man of exceptional knowledge
ıd ability, who was instrumental in
ıilding the Smithsonian Institution
to one of the greatest museums of
e world.

Above Right. Elliott Coues, doctor,
rnithologist and prolific author
ho pioneered the exploration of
ıe avifauna of the U.S.A.

). Henry Seebohm, Victorian businessman,
ho travelled widely throughout
urope and Asia collecting birds
nd eggs.

11. Harry Witherby, whose work as author and publisher was a major factor in the development of ornithology in Britain in the first half of this century. Shown here with F.C.R. Jourdain (Left).

2 & 13. Witherby's Handbook established the concept of a detailed handbook which is still a valuable tool for ornithologists. It was preceded by the rare Practical Handbook of 1920 which is shown top centre of plate 13. More compact successors are the two Popular Handbooks by P.A.D.Hollom. When The Handbook Of The Birds Of Europe is completed in seven volumes it will be the best single information source for birds of The Western Palearctic, a fit successor to Witherby's concept.

14. Lord Walter Rothschild, an exceptional naturalist who built up a great Museum from his own financial resources.

15. John Gould, the originator and publisher of a whole series of magnificent bird books.

16. Richard Kearton was a pioneer bird photograph who also wrote and illustrated many books.

. Above Left. Roger Tory Peterson's
ame is synonymous with the
odern field guide.

. Above Right. David Lack, particularly
otable for his outstanding work in
e field of bird behaviour and
opulation studies.

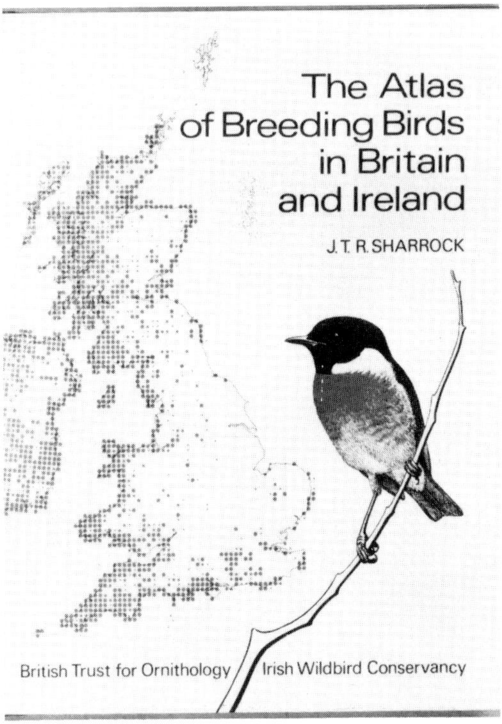

. The BTO Breeding Bird Atlas which
arked the advent of the era of
etailed distribution studies.

20. The concept of field guide, originated with the Peterson publication of 1934 which has been constantly updated. Field guides are available for almost all parts of the world, the range shown above being a very small selection.

Walter Rothschild. Brilliant, extravagant and wildly eccentric he occupies a unique position in the annals of British zoology.

He was born in London in February 1868 and educated, on the insistence of a doting mother, at home. Later he travelled for a while in Germany spending some months at Bonn University before going up to Magdalene College, Cambridge in 1887. He did not complete his course of studies there and left without taking a degree. From earliest childhood natural history was his exclusive and abiding passion, and by the age of seven he was already dreaming of having his own museum — an ambition he had achieved by the age of ten, when he started a small museum of stuffed birds in a shed at the end of a garden in Albert Street, Tring. This was ultimately to expand into his "real" museum, the building of which commenced in 1889 when he was twenty-one. This was eventually to contain 2,000 mounted mammals, 1,400 mammal skins, 2,400 mounted birds plus 30,000 bird skins, 200,000 birds eggs, 680 reptiles plus 300 dried reptiles, 914 fish, two and a quarter million lepidoptera and some 30,000 beetles: there was also a library containing some 30,000 books. Lord Rothschild remained an obsessive collector throughout his life, and spent virtually the whole of his vast personal fortune on enlarging and improving his museum. Morever it was never intended that this should be a treasure house the contents of which were to be enjoyed exclusively by himself or his friends, but rather a lively and active working research establishment which disseminated its findings in some forty volumes of "Novitates Zoologicae", and published such key works as "Die Vogel der Palaarktishen Fauna" and the "Revision of the Lepidopterous Family Sphingidae".

Walter Rothschild was an extremely shy man who had the misfortune to suffer from a difficult speech impediment, this only disappeared when he was talking about something which really interested him when he could be quite eloquent. In many ways he was a strange and complex man with a somewhat unworldly attitude to life, particularly where money was concerned.

He was always very close to his mother, whose favourite he was and who spoilt him throughout his life. The psychological aspects of his personal relationships do not concern us here, with the exception of one mysterious affair which ultimately led to his being blackmailed by a peeress: this is of importance in that it was one of the factors which forced him, in 1932, to sell the most valuable part of his bird collection, including priceless and irreplaceable skins of extinct birds, to the American Museum in New York. These blackmail demands however were not the sole cause of his cash crisis; the extravagant purchases for his museum and the cost of maintaining a world-wide network of agents and collectors (estimated during his most active period around 1890 to 1908 at about four hundred) strained even his resources. The work of these collectors was planned and organised by him in the minutest detail, he possessed a quite remarkable memory and could recall exactly where each collector was at a given time and the exact instructions they had been given.

One of the most important expeditions he mounted was to the Galapagos Islands, the primary object of which was to bring back some giant tortoises in which he was fanatically interested. He had a great collection of birds made of the Sandwich Islands (Hawaii) and Laysan Island, including many which were extremely rare and later became extinct. The results of this expedition formed the basis of one of his most important bird books entitled "The Avifauna of Laysan and the Neighbouring Islands, with a complete history to date of the birds of the Hawaiian possessions". This was issued in three parts between 1893 and 1900 with 55 coloured plates by Keulemans and Frohawk.

Another important bird book by Lord Rothschild is "Extinct Birds" published with 45 coloured plates in 1907; the artists were Keulemans, Lodge, Gronvold, Smit and Frohawk. This was a most useful and innovative book in that it contained illustrations of several species which were not to be found elsewhere at that time. An important work by Lord Rothschild, from the scientific point of view, was "A Monograph of the

Genus Casuarius'' containing 18 coloured plates and published in 1900. This was not produced separately as a book being a part of ''The Transactions of the Zoological Society of London'' for that year; there are however a few copies in existence which were bound up separately.

Lord Rothschild died in August 1937. With a minimum of formal training he had succeeded in becoming a recognised world authority of not one but several branches of zoological study. For his drive and energy, his work in organising and financing expeditions and the work of others, British ornithology must be forever in his debt. His museum at Tring, together with its contents, he bequeathed to the nation. It would have pleased him to know that today it remains open to the public throughout the year and receives tens of thousands of visitors from all over the world: it also, most appropriately, houses the sub-department of Ornithology of the British Museum, (Nat. Hist.).

VIII

The Emphasis On Bird Behaviour

Until the early years of this century the principal aim and purpose of ornithology and its literature had been directed mainly towards the listing, distribution and identification of birds and their eggs. Another objective, and one which had also attracted a fair degree of attention from early ornithological days, was the study of the anatomy of birds. There were undeniably sound reasons for the emphasis placed on the importance of these particular lines of study: firstly, there was a very great deal to be discovered within these specific areas and, secondly, up until this time there had been a pronounced lack of any of the practical equipment so necessary for the study of the movements and habits of a living as distinct from a dead bird. Binoculars and lightweight telescopes were not generally available until the latter part of the nineteenth century, making it practically impossible to observe properly and accurately such activities as nest building, display, feeding and migration etc. The general ignorance about such subjects was compounded, so far as the smaller passerines such as the warblers, finches and the like were concerned, by the fact that small, not very colourful birds aroused little interest in the Victorian and Edwardian ornithologist, the majority of whom preferred to concentrate their attention upon the birds of prey, game birds and the more exotic and flamboyant species. "Why doesn't he spend his time finding the nest of his strange 'eagles' intead of wasting his time on small birds?", enquired F.C.R. Jourdain in a letter written to Horace Alexander, concerning his brother Christopher, just before the first world war. This indifference was particularly marked in the case of birds in intermediate plumage with all the accompanying difficulties of identification: on the whole it remained easier to follow the general rule and if in doubt shoot it, for, as the adage went, "What's hit's history, what's missed mystery".

From 1900 onwards the movement away from classification and collecting as the primary objectives of ornithology began to gather pace, and that of behavioural study or ethology started gradually to assume some of the importance which has been acquired in research today. The true pioneer of ethology in Britain was Edmund Selous, who commenced his immensely detailed accounts of the activities of birds at some time during the 1890's, although his first paper on the subject was not published until 1899 in the periodical "The Zoologist". Largely unappreciated and unread at the time, it is only comparatively recently that the full value and importance of his work has received its proper recognition. Writing in 1959 on the occasion of the centenary of the foundation of the British Ornithologists Union, David Lack placed Selous' book "Realities of Bird Life" as being his own first choice amongst the five most important bird books which had been written in Britain between 1899 and 1939. This decision, must have caused some surprise to some of his readers, many of whom had probably never heard of Edmund Selous, let alone read any of his work.

In view of his strange, lonely and withdrawn character, it is hardly surprising that so little is known about Edmund Selous. Unlike his elder brother, Frederick Courteney Selous, a noted big game hunter and naturalist, his work received so little recognition during his lifetime that after his death no British journal published either an obituary or any appreciation of his life and work. The two obituaries which did appear in "Alauda" in France and "Bird Banding" in the U.S.A. do not enlighten us much further on this enigmatic and shadowy figure. Writing in "Alauda", Jacques Delamain simply commented that, "Selous was a genius, lonely, misunderstood, unappreciated, a tragic figure in his old age, with his passion for the lovely wild things of the earth that are relentlessy driven to their doom by cruel and indifferent man''. During the half century which has elapsed since his death, despite some serious research, very little fresh knowledge about him has emerged, and as recently as October 1984 there was an appeal in ''The Ibis'' asking for any information regarding Selous' articles published in non-scientific journals and for any further biographical data.

Edmund Selous was born Edmund Slous in 1857. Little is known of his early life except that in 1877 he went up to Pembroke College, Cambridge, in order to read law and was called to the bar in 1881. He practised as a lawyer until 1888 when he went with his wife and young family to live in Wiesbaden in Germany remaining there until 1891. As a result of this time spent in Germany he and his family invariably conversed in German when amongst themselves. On his return to England he went to live at Wamil Hall near Mildenhall in Suffolk, later moving to Icklingham in the same county. Here, situated near the Brecklands, he devoted his entire time and energy to the study of bird behaviour, walking or cycling many miles in that ornithologically rich area, and patiently spending hours watching and making voluminous notes, written in an untidy scrawl and with no semblance of order. These notebooks, now in the Edward Grey Institute in Oxford, he called his ''observational diaries''. They contain much valuable and hitherto unrecorded pieces of information — ''nuggets'' as James Fisher called them — but like golden nuggets they are embedded and scattered amongst a mass of extraneous material.

His first behavioural study on the nightjar was published in 1899 and this was followed by a whole series: ''The Stone Curlew'' (1910), ''Great Crested Grebe'' (1901), ''The Black Grouse'' (1909-1910), ''The Sparrowhawk'' (1911), ''The Red-throated Diver'' and ''Carrion Crow'' (1912), all of which appeared in ''The Zoologist''. When this periodical ceased publication in 1916 he was tremendously upset and absolutely refused to write anything for its successor ''British Birds''. Instead he wrote a few articles for a new but short-lived magazine "Wildlife" which ran from 1913-1918. These articles consisted of detailed studies of the behaviour of the little grebe and the shag.

Selous's books fall into two entirely different and separate categories. Those written for children such as "Tommy Smith's Animals" (1899), "Beautiful Birds" (1901), Tommy Smith's Birds" (1922) were produced mainly to augment his small private income, but the somewhat "twee" anthropomorphic avian dialogue masks some fairly serious behavioural studies. His six serious behavioural books with their long convoluted sentences make for heavy reading today, but at the time they were published they were unique both as to subject and content, and are all still of great interest. The first, "Birdwatching" was published in 1901 being in part a reprint of some of his articles which had already appeared in "The Zoologist". "The Birdwatcher in the Shetlands" (1905) contains important comments concerning the display of seabirds as well as observations upon seals. The same year saw the production of "Bird Life Glimpses" which included many of his observations and discoveries carried out in Suffolk. There followed a lengthy hiatus before, "Realities of Bird Life" which was published in 1927.

His next two books were "Thought Transference (Or What?) in Birds", and another highly important publication, "Evolution of Habit in Birds": the latter which contains much thought provoking material, is immensely difficult to read and calls for much patience and concentration on the reader's part if he is to obtain the full benefit of Selous' thoughts and conclusions. Selous left behind him a seventh and unpublished work, "Alone with Birds in Iceland".

Although it was not until long after his death that Selous came to be recognised for the brilliant and innovative ornithologist that he undoubtedly was, there were even during his lifetime a few exceptions to this general indifference. In 1927 in an introduction to "Realities of Bird Life", Sir Julian Huxley wrote: "The general impression made by most books in those days on a young bird-lover was that British ornithologists had nothing much left to discover. The books were full of facts, but did not trouble themselves to enter details of habits, nor to point out the gaps in knowledge. Today all this is altered, the modern ornithologist and bird lover has come to realise that he has only arrived at a first approximation to knowledge". It was precisely these gaps mentioned by Huxley that Edmund Selous through his writing involved and ponderous though they often were, strove so hard to fill.

Whilst it is regrettable that so little is known about Selous it is even more deplorable that during his lifetime he received no recognition or honour for his work from any scientific body. So completely were his writings undervalued or ignored by the ornithological world that none of his books was ever reviewed either in "The Ibis" or "British Birds", nor, as far as we know, was he ever aware of the increasing notice and esteem accorded to his work by a few of the more perceptive of his contemporaries. During his last years he became an embittered recluse, refusing to read any ornithological publications. He died, lonely and unknown, in Weymouth in March 1934, a strange and shadowy figure until the end.

An excellent observer of avian behaviour and a contemporary of Selous', although entirely different in character and outlook, was Eliot Howard. Howard's studies were chiefly directed towards trying to analyse the mental processes of the bird rather than giving a direct description of the bird's actions, a line of research which led him into some deeply involved speculation. Despite these excursions into the realms of metaphysics, the works of Eliot Howard are of considerable value, particularly as regards bird behaviour, although, like Selous, much of his writing is difficult to follow and understand.

His first book "The British Warblers, a History with Problems of their Lives", published in ten parts between 1907 and 1915 was a beautiful book. It was illustrated by Gronvold and contained 35 colour plates besides some 50 plates in black and white. This book is noteworthy in that it was, on several scores, a pioneer work. As well as being the first serious study of a family of "small brown birds", it was the first monograph to concentrate on the behaviour of the bird rather that its physical character and distribution, besides containing much original observation on courtship and territory. Howard's observations and interpretation of courtship and territory was set out in even greater detail in his second book "Territory in Bird Life" (1920). In "The Nature of a Birds World" published in 1935 he attempted to set out the relationship between the physiology and the psychology of avian breeding cycle. This again was an original and pioneering effort despite pursuing a somewhat impracticable line of thought difficult to develop: since it is impossible to judge by what emotional processes the actions of a bird are governed it follows that it is equally impossible for any such conclusions to be tested or proved. "Too few obervations and too many speculations" was a pretty fair criticism of the book, and many people felt that Howard was too heavily influenced by Lloyd Morgan, the Professor of Psychology at Bristol.

Eliot Howard was born at Kidderminister in 1873 and was educated at Eton and Maran's College, Birmingham. An eminent industrialist, most of his bird study was confined to his spare time, whether in his own garden and the countryside around his house which overlooked the river Severn, or on visits to the coast. Although his work was well known to his contemporaries, he was a quiet and retiring man, so much so that a book was "dedicated to his memory" some twelve years before his death, which actually occurred in 1940.

Prior to 1900 bird migration, like their display and behaviour was a subject about which very little was known for certain. The theory of migration had been widely accepted since Gilbert White's time, but little serious work towards its amplification or verification was carried out until the 1880s, when the British Association for the Advancement of Science appointed a special committee to study bird migration, a world pioneering effort in this field. Later, in 1904, the British Ornithologists' Club formed another committee with the object of collecting records from around the country registering the arrival and departure dates of the summer visitors. These results were published in nine annual reports. The pilot study for the British Association's scheme was carried out by John Harvie Brown and John Cordeaux, with the assistance of William Eagle Clarke. The final interpretation, written at the suggestion of Professor Alfred Newton, was made by Eagle Clarke in his classic work, "Studies in Bird Migration", published in two volumes in 1912. This book, along with those of Selous and Eliot Howard was included in David Lack's list of the five most important works to be published in Britain since 1899.

Owing to their geographical position the British Isles receives, during the course of the year, birds from almost every direction: they do not however, except on very rare occasions, experience the visible mass migrations which occur in such places as the Bosphorus, Falsterbo or Gibraltar. So in order to study migration systematically, it was necessary that a large number of observation points should be established all around the United Kingdom, mainly in the coastal areas.

In preparation for collating the reports from these widely scattered sites and incorporating them into his book, Eagle Clark considered carefully the many coastal areas likely to be the most suitable from where to observe migration, and amongst these was Fair Isle, lying between the Orkneys and Shetland. He was the first person to appreciate its peculiarly advantageous position both as an ornithological observatory and a locality on which large numbers of rare species regularly occur. He visited the island frequently as well as employing a permanent observer there. The man appointed for this job was George Stout, who proved to be an excellent choice, supplying much valuable information concerning the birds of Fair Isle and their migration until his untimely death, on active service, during the first World War.

William Eagle Clarke was born in Leeds in 1850 and educated at Leeds Grammar School, and Yorkshire College (later to become Leeds University). On completing his education he first worked for a firm of civil engineers, in 1879 he published his first scientific paper on the study of birds and in 1884 was appointed curator of Leeds Museum. From there he went on to Edinburgh in 1888 as an assistant in the Natural History Department of the Royal Scottish Museum, becoming the Keeper of the Museum in 1906.

His contribution to ornithology was notable not only for its originality but also for its accuracy and the clear and orderly nature of his reports and discoveries. He was honoured by the British Ornithologist Union in 1922 by the award of the Godman-Salvin medal, and was elected to be the Union's President in 1918. Besides being an excellent all round naturalist he was a keen fisherman and a great walker who was often to be seen striding over the Pentland Hills, or spending hours watching the wildfowl and waders at Aberlady Bay. In the early thirties his health began to fail and, following a long illness, he died in May 1938.

Another ornithologist from whose work the study of migration benefited greatly was Richard M. Barrington, whose ingenious idea it was to collect records from the keepers of the Irish lighthouses and publish the results in a book with the self-explanatory title of "The Migration of Birds as observed at Irish Lighthouses and Lightships" (1900). This exceedingly detailed work helped to give a much clearer picture of migration as it occurrred around Ireland, an area which had hitherto been very sparsely covered by other sources. An interesting appendix to this book gives the wing measurements of some thousand dead birds, and was probably the first large sample of such measurements ever to be published.

The practice of placing rings on the legs of birds in order that they could be positively recognised if recovered at a later date away from the ringing area made it possible, for the first time, for the birds' routes to be plotted with some degree of accuracy: this, in turn, led to a revolution in the knowledge and understanding of bird migration and hence all the literature on the subject. The first two ringing schemes in Britain were both started in 1909, one being organised by the magazine "British Birds" founded by H.F. Witherby, and the other at Aberdeen

University at the instigation of Arthur Landsborough Thomson. The "British Birds" scheme was the larger of the two and, in 1937, was handed over to the newly founded British Trust for Ornithology to become a part of the national scheme which still flourishes today.

Arthur Landsborough Thomson's important contribution to the study of bird migration during its early and formative years was not the only branch of ornithological study for which he is remembered; and although technically he was an amateur his background and approach to this subject were wholly scientific. Born in Edinburgh in 1890, he was the son of the Regius Professor of Natural History at Aberdeen University, which, in due course, he himself attended. Following his graduation he went on to do postgraduate work at the Universities of Heidelburg and Vienna. His first ornithological excursion of importance was in 1907, when he spent a fortnight at the famous Prussian bird observatory at Rossiten (now Rybachy) on the Baltic coast. Here, for the first time, he encountered the practice of bird ringing, and was so impressed that he determined that a similiar scheme must be organised in Britain. This work was interrupted by the first World War, during which he served with the Argyll and Sutherland Highlanders until he was wounded in 1916. Following his demobilisation in 1919 he started his long career with the Medical Research council, which today owes much to his clear vision and superb organising ability.

Landsborough Thomson's work for ornithology falls into three categories: his pioneering work on the study of migration, his ability to organise and combine the efforts of the many and diverse scientific bodies of which he was a member, and thirdly, and most importantly, his books. The first of these was a popular work entitled "Britain's Birds and their Nests Described" (1910), of rather more importance were his later books on migration, "Problems of Bird Migration" (1926) and a revised work "Bird Migration, a Short Account" (1942). His greatest work, and one which took him eight years to accomplish, was his editorship of "The New

Dictionary of Birds" published in 1964, and to which he himself contributed forty entries. This huge encyclopedia of ornithology containing some half a million words was produced in celebration of the centenary of the British Ornithologists Union, (which actually occurred in 1959).

The enormous and widespread influence which he wielded over British ornithology for so many years make Landsborough Thomson one of the most noteworthy contributors to the science during the 20th century. Amongst the scientific bodies of which he was a member was the Council of the Zoological Society, the Council of the Fauna Preservation Society, Chairman of the Home Office Advisory Committee for the Protection of Birds and Chairman of the Trustees of the British Museum (Nat. Hist.). During his eight years as President of the British Ornithologists Union he carried out a complete financial reorganisation of that body. Honours were showered on him both at home and abroad, honorary doctorates, the Buchanan Medal of the Royal Society, the Godman-Salvin medal of the British Ornithologists Union, the Bernard Tucker medal for the British Trust for Ornithology, Membership of the Netherland Order of the Golden Ark, and the C.B. in Britain. Despite these honours and recognition he remained a quiet and unassuming man, possessed of a great sense of humour and a fine dry wit. He was sadly missed after his death in 1977.

Although originally it had been British ornithologists who had been the forerunners in the study of bird behaviour, by the 1930's the lead in this field had passed to workers in Holland and Austria. One of the foremost Dutch behaviourists whose work has contributed notably to this particular line of study was Nikolaas (Niko) Tinbergen, although his research, especially that on the behaviour of the stickleback, has gone far beyond the bounds of ornithology.

Born in 1907, he read Zoology at the University of Leiden. Like Landsborough Thomson before him, Tinbergen's interest in this subject was fired by a visit to the same bird observatory at Rossitten, and he was also greatly influenced by the writings of Oskar Heinroth and Julian Huxley. In 1932-33 he spent a year in east Greenland, and as a result of this expedition wrote "The Behaviour of the Red-necked Phalarope in Spring", which appeared in the Dutch journal "Ardea" in 1935. This was followed by "The Behaviour of the Snow Bunting in Spring" published in the "Transactions of the Linnaean Society of New York" in 1939. During the thirties he established a close friendship with another leading behaviourist, Konrad Lorenz, and the two men worked together in Austria for a short time. The war put an end to all such research, and it was not until 1949 that Niko Tinbergen came to England to take up a lectureship at Oxford University.

Soon after his arrival in Britain Tinbergen wrote two extremely important books, "Study of Instinct" (1951) and "Social Behaviour of Animals" (1953). Although neither of these books were devoted exclusively to the study of bird behaviour they were key works in the development of the science of ethology — the biological study of animal behaviour. "The Herring Gull's World, a study of the Social Behaviour of Birds" (1953) and a very popular work "Bird Life" (1954), although published in England were the fruit of field work carried out by mostly in Holland mainly at Wassenaar and Terschelling just after the war. Unlike Selous' work, "The Herring Gull's World" is precise, lucid and easily readable, illustrated with excellent photographs and line drawings, it demonstrated very clearly the great advances which had been made in ornithology during the previous fifty years.

A remarkably able lecturer, Niko Tinbergen's influence was by no means confined to the undergraduates at Oxford. During the fifties one of his original and innovative ideas was to set up field camps at which he introduced young naturalists, drawn from all around Britain, to his ideas and concepts concerning behavioural study. Two of the sites he chose were the Farne Islands in Northumberland and Ravenglass in Cumbria, where he himself did much work on the black-headed gull colonies.

In 1966 he became professor of Animal Behaviour at Oxford University, and the following year produced a highly interesting book called "Tracks", which he wrote in collaboration with Eric Ennion. By cleverly combining photographs and drawings they showed how, by means of footprints, wing marks in the sand and other signs, the behaviour of the colony could be interpreted and understood. This technique was repeated and further developed by Tinbergen and Ennion, this time in collaboration with Hugh Falkus, in "Signals for Survival" (1970): which was based on a prizewinning television film concerning the display and behaviour of the lesser black-backed gull. This method is an admirable way of explaining to the amateur the intricacies of patterns of behaviour in a familiar and easily observed bird.

Professor Tinbergen's contribution to the progress of ethology in Britain can hardly be exaggerated, and it is mainly due to his guidance and example that a new generation of scientists have continued to widen and advance this particular scientific discipline.

Much of Tinbergen's early work was published in Dutch and therefore falls outside the scope of this book. A similiar restriction also applies to the written work of his fellow pioneer, Konrad Lorenz. Born in Vienna in 1903, he studied biology and medicine before becoming a professional zoologist, concentrating his research on the behavioural studies of birds and other animals. Prior to the second world war his work was published entirely in German, but a popular book "King Solomon's Ring" (1952) was translated into English and has since become something of a classic. Using fish and mammals as well as birds as his examples, this was written in a light style with amusing line drawings, giving the reader a clear outline of the science of ethology. Konrad Lorenz's avian studies have been largely devoted to the greylag goose, a species about whose behaviour he is the world's greatest living authority. In 1979 he published "The Year of the Greylag Goose" which was illustrated with excellent colour photographs by Sybille and Klaus Kalas. Although not a scientific work in the strictest sense this was a serious study written in non-technical language.

Throughout its history ornithology has been studded with enthusiasts who have had a particular obsession with one particular bird: for Professor Newton it was the great auk, for James Fisher the fulmar, for Edgar Chance it was, above all, the cuckoo. Born in 1881 at Edgbaston, Birmingham, Edgar Chance went up to Trinity College, Cambridge, before commencing his business career. He was not a trained scientist but an amateur ornithologist whose interest lay primarily in egg collecting. As a collector he became fascinated by the degree of mimicry of which the cuckoo was capable and its remarkable ability in matching the colour and pattern of its eggs with those of its host.

The efficient and business-like methods which he brought to his studies of the cuckoo and later the red-backed shrike made them models of well-planned and executed research. A superb observer himself, he organised several other competent observers to assist him, amongst whom was Oliver Pike, a pioneer nature cinephotographer. From careful observation Chance knew that in his chosen study area, Pound Green Common, Upper Arley near the Wyre forest, the cuckoos were parasitising meadow pipit's, and accordingly he and his helpers found every meadow pipits nest within that area and commenced watching the female cuckoos constantly throughout the daylight hours when they were laying their eggs. With experience the watchers found that they could predict when and at what nest a particular cuckoo would lay. Previously there had been numerous theories as to how a female cuckoo found a potential host nest and laid her egg into it, but through his remarkable piece of research Chance was able to settle these questions decisively.

The book which recorded this work, "The Cuckoo's Secret" was published in 1922, and was a valuable addition to behavioural studies. An updated version, "The Truth about the Cuckoo" appeared in 1940, but by this time such was the dislike and suspicion in which egg collectors like Chance were held that he received little recognition. Following his prosecution for taking crossbills eggs in East Anglia he was forced to resign from the British Ornithologists Union, his death in 1955 was not reported in "The Ibis" and barely noted in "British Birds". However quite regardless of the ethics of egg collecting, Edgar Chance's contribution to ornithology was far from negligible and should be properly acknowledged.

The serious study of bird behaviour being carried out in Europe during the early part of this century was paralleled in the United States in the work of Margaret Morse Nice. Born Margaret Morse in December 1883 in Amherst, Massachusetts, she graduated from Mount Holyoke College in 1906, a year later than intended because she took a year off to travel in Europe in order to improve her foreign languages. That same year she went on to Clark University at Worcester where, in 1909, she married a fellow graduate. Her particular interest in birds started at some time around 1907 and in 1910 she published a paper on "The Food of the Bob-white", for which she finally received her doctorate in 1915.

In 1913 she and her husband moved to Oklahoma and it was during her time here that she began her serious behavioural studies which continued when the family moved to Columbus, Ohio in 1927. In 1924, in collaboration with her husband Leonard Blaine Morse she wrote her first book, "The Birds of Oklahoma". In the following years she published numerous papers in "The Condor", "Wilson Bulletin" and "Bird Lore", besides writing the section on "The Theory of Territorialism and its Development" in the American Ornithologist Union book which was published in celebration of its 50th anniversary in 1933. In the same year her first major paper on the song sparrow appeared in the "Journal fur Ornithologie", entitled "Zur Naturgeschichte des Singammers" published in Germany at the request of Dr. Erwin Stresemann.

Following her move to Chicago in 1936 she continued her work on the behaviour of the song sparrow, and it was at this period of her life that much of her field work was done, including a great deal of ringing and weighing of the birds within her study area. The "Life History of the Song Sparrow I" had already appeared in the "Transactions of the Linnean Society of New York" in 1933, she continued to expand this particular study and "Life History of the Song Sparrow II" was published in the same journal in 1943. Both papers were eventually produced in 1964 under the title of "Studies of the Life History of the Song Sparrow", in two volumes in paperback.

In all Margaret Morse published more than 250 papers in various scientific journals, some of them in German, and a few of which concerned European birds such as the golden oriole. In 1939 "The Watcher at the Nest" was published, this was a popular work containing some of the material from her research on the song sparrow, as well as chapters on herons, mourning doves and bobwhites. An excellent reviewer she contributed over 3,300 articles to various scientific publications. She was a painstaking and meticulous correspondent with other workers in her field, often displaying great forbearance with those who contacted her on trivial ornithological matters. As she grew older however she became extremely outspoken and many eminent people, including the President of the United States, were bombarded with her strictures and admonishments concerning the use of pesticides and the misuse of wildlife reserves.

Whilst she was not, properly speaking, a professional in the strict sense of the world, she had received a thoroughly professional training. As she herself said on more than one occasion, "I am not a housewife, I am a trained zoologist". As she was also a devoted wife and mother her husband and children would probably not have agreed with the first part of this remark!

Margaret Morse Nice received many tokens of recognition from ornithological organisations throughout the world. She was elected to honorary membership of the British Ornithologists' Union, and to ornithological societies in Germany, Finland, the Netherlands and Switzerland. She was President of the Wilson Ornithological Society in 1938 and 1939 and associate editor of the Wilson Bulletin for ten years.

In 1942 in acknowledgement of her work on the song sparrow she received the Brewster Medal of the American Ornithologists' Union. She died at the age of ninety in June 1974.

Finally, although the main emphasis of his work lay outside the ornithological field, no review of behavioural studies can ignore the contribution of Sir Julian Huxley, Professor of Zoology at Oxford from 1919-1925 and later at King's College, London. Amongst the numerous papers which he published in various scientific journals were two on the courtship of the red throated diver (1914) and the great crested grebe. The latter was reprinted separately as a small book in 1968 under the title of "The Courtship Habits of the Great Crested Grebe". By virtue of his name, reputation and academic status, Huxley was able to convince the scientific establishment that far from being a mere hobby, ornithology was a respectable subject worthy of serious study in its own right.

IX

The First Popularisers

Although a general interest in all the aspects of natural history had been steadily growing throughout the 19th century, its various branches were highly susceptible to the vagaries of fashion. Botany, lichens and seaweeds were amongst the diverse subjects which were enthusiastically "taken up" at different times. Acutely aware of this shifting market, the publishers were quick to choose their titles accordingly. It was inevitable that in due course ornithology should take its turn, although it never acquired in Victorian times the mass popularity which it enjoys today. In his book "The Naturalist in Britain" D.E. Allen suggests that it was because "Ornithology had long been accepted as an extension of field sports which enabled it to partake of some of their atmosphere and high social standing". This is a very likely explanation, for certainly as far as the British Ornithologists' Union was concerned the list of members during the first decades of its existence reads rather like "Kelly's Directory of the Landed Gentry" laced with a liberal admixture of "Debretts". Until the 1930s only men could be admitted to it as full members, whilst (perhaps in retaliation) for several years after its formation The Society for the Protection of Birds admitted only women.

From 1832-1861 the population of Britain rose by 40%, but during the same period the total number of children attending day schools rose to 68%. With increasing literacy there arose an unprecedented demand for all kinds of books from cheap 'yellow' fiction to those of an 'improving' nature. Amongst the new outlets required for the sale of these publications were the numerous bookstalls, the first of which opened at Euston Station in London in 1848.

Whilst a good deal of trivia was produced at this time much was of a consistently reasonable quality, and this was particularly true of the many ornithological publications which now began to arrive on the market: Walter Swaysland's "Familiar Wild Birds" (1883) produced in four volumes with coloured plates by Archibald Thorburn, and "Sketch Book of British Birds" by Richard Bowdler Sharpe (1898) were fairly representative examples. A prolific author who was very much to the fore in writing the texts for cheap popular bird books (generally illustrated by printed colour plates) was the Reverend Charles Alexander Johns. "British Birds in their Haunts" was the first published in 1862 at a price of 12/- (60p) and was still in print in 1948, having passed through numerous editions from many different publishers. The illustrations in the first edition were drawn by J. Wolf and engraved on wood by Whymper. A later edition revised and annotated by J.A. Owen depicted 256 birds on 64 coloured plates drawn by W. Foster. The 1948 edition which cost 16/- (80p) contained an additional five black and white plates by Roland Green, the text being amended and edited by W.B. Alexander. Much of Johns' work such as "Flowers of the Field" and "Home Walks and Holiday Rambles" (1863) are dismissed today as simply being 'pot-boilers', which, in a sense, is true; but for a great many people who wished to know more about these subjects his books, reasonably priced and containing good coloured illustrations, were far easier to use for bird identification than the detailed descriptions and black and white engravings as supplied in Yarrall's or Saunders' works.

Advances in printing and increased demand had a dramatic effect in reducing book production costs.

The Rev. Johns' "British Birds in Their Haunts" long printing history with the figures given above invites interesting comparison with people's income during the eighty or so years the book was in print. The conclusion must be that books have become remarkably cheap over the years. Because of the cost advantage photography was to displace the artist, though only for black and white illustrations. Colour photography was not an economical process until well into the twentieth century.

The first bird book in Britain to be illustrated by photographs was "The Birds of Berkshire and Buckinghamshire, a contribution to the natural history of the two counties", by Alexander William Maxwell Clark Kennedy, this was published in 1868 whilst he was still a sixteen year old schoolboy at Eton and contained four very poor views of stuffed birds. It was another twenty years before cameras were devised using stable plates which could be handled easily and not require the immediate use of a dark room, so that practical field photography was possible.

Two of the most eminent pioneers in bird photography in Britain were Richard and Cherry Kearton, the sons of a sheep farmer in Swaledale in Yorkshire. Throughout their long and productive lives the two brothers worked closely together on numerous publications, Richard generally as the author and Cherry as the photographer. Their first book "British Bird Nests, How, Where and When to Find and Identify Them" was published in 1895 at a price of one guinea (£1.05p) and was illustrated with many photographs.

Many of their works such as "Strange Adventures in Dicky Bird Land", "Stories told by Mother Birds To Amuse Their Chicks" (1901) and "The Adventures of Cock Robin and his Mate" (1904) were produced expressly for children who, because of a steadily rising birth rate and universal literacy, now formed an increasingly large part of the booksellers' market. A more serious and adult book "Our Rarer British Breeding Birds, their Nests, Eggs and Summer Haunts" was published in 1899, and although many of the birds included in it such as redshank, redpoll, hawfinch and blackcap were in fact far from rare, the photographs were of fair quality albeit still mostly of nests. "British Birds' Nests" published in 1907 made use of some of the better photographs from the earlier book whilst containing many new photographs of the birds themselves, a few of them actually shown in flight, besides including a number of colour plates of eggs. The illustrations were far better than in their earlier works and showed clearly how the Keartons were developing their methods and techniques in response to the advances and improvements in photographic equipment.

Richard was a particulary good field observer, and because of the limitations of contemporary cameras he was forced to spend long periods quietly watching and waiting; his resulting observations of the behaviour and breeding strategy of a large variety of species were of considerable value. Their small book "Our Bird Friends" published in 1900, although rather heavily slanted towards eggs and nests, was a useful introduction to birds for young people. Whilst it is easy to criticise this bias towards nests and eggs, the exceedingly heavy plate cameras and the slow film speed available to photographers at this time, meant that most opportunities of photographing birds were mainly confined to the nest and its immediate surroundings. The Keartons showed great ingenuity in their use of hides, from constructing artificial tree trunks to their famous stuffed bullock in which one of them would conceal himself complete with camera. Concerning this last device there is in existence a wonderful photograph of an incident when Richard, overcome by heat and cramp, overbalanced and the bullock is shown lying on its back with its four legs in the air, plus two human legs as well!

In 1882 Richard Kearton moved from Yorkshire to London to become the submanager of the publicity department of Cassells, the publishers; and following the death of their father Cherry came to live in London also. Determined to obtain new and exciting pictures they both travelled widely, in the course of which they endured a good deal of discomfort from standing for hours in water up to their chests, to hanging on ropes over the great bird cliffs of St. Kilda and Ailsa Craig.

In order to extend his range of photographic studies of wild life Cherry travelled to Kenya and also visited India in 1911. Following this he went to North America and then in 1913 returned to Africa. Much of his later work in Africa involved cine photography and is outside the scope of this book.

The Kearton brothers' work marked the point at which

books, formerly always illustrated by artists, became superseded, especially at the lower end of the market, by photographic illustrations.

Richard Kearton died in 1928 and Cherry in 1940. Although they were self-taught and not professional ornithologists nevertheless through the widespread popularity of their books they played a considerable part in both promoting and widening an interest in birds amongst the general public in Britain.

Another early bird photographer was Reginald Badham Lodge — the elder brother of George Lodge the famous bird artist. He was born at Louth in Lincolnshire in 1852 and in due course set up a portrait photographer's business at Enfield in Middlesex. Being a professional photographer and a keen ornithologist he was naturally very interested in photography in the wild, and to this end he developed a number of modifications and improvements to his camera which enabled him to take better photographs outside the environment of his studio. Amongst these inventions was an automatic shutter release which allowed the birds to photograph themselves, and in 1901 he produced a reliable electric shutter which was released by the weight of the bird's foot when it alighted on a particular perch. All this equipment was still both heavy and awkward to move around, so much so that at one stage he used a wheelbarrow to carry his 12" x 10" plate camera from place to place.

Reginald Lodge travelled widely on photographic trips, visiting Holland, Spain, Denmark and the Balkans, and his book "Bird Hunting through Wild Europe with a Camera", published in 1908, gives an excellent account of these excursions. Like the Keartons Lodge did not make any great direct contribution to ornithological knowledge, but through his photographs he helped to stimulate the increasing public interest in and awareness of birds.

A photographer who was much influenced by Lodge, at least in his younger days, was Oliver G. Pike. Born in 1877 he met Lodge sometime around 1893 and helped him on some of his photographic expeditions. He soon became an expert photographer in his own right, taking his first pictures of garden birds in 1895. Deciding that the equipment available to him was unsuitable for his particular needs he designed his own camera of the reflex type which was sufficiently portable for use when stalking birds, this camera he proceeded to market under the name of "The Birdland Camera". At much the same time similiar cameras were being developed, quite independently, by Voigtlander in Germany and in the U.S.A., all of them, by today's standards, still extremely bulky.

Oliver Pike embarked upon a whole series of popular books including, "In Birdland with Field Glass and Camera" (1900), "Woodland Field and Shore — Wild Nature Depicted With Pen And Camera" (1901), "Hillside, Rock and Dale" (1902), "Home Life in Birdland" (1905), Birdland Pictures" (1906), "Adventures in Bird-land"' (1907) and "Behind the Veil in Birdland" (1909). He was a pioneer in the use of cinematography, and as early as 1913 shot a remarkable film showing a nestling cuckoo evicting a baby reed warbler from the nest. His expertise and interest in the cuckoo led to his meeting and collaboration with Edgar Chance, for whom he was able to provide photographic evidence of Chance's theory as to how the hen cuckoo laid her egg in the host nest. Oliver Pike spent two months working with Chance. He was able to employ a new development of high speed photography, which clearly disclosed many fresh revelations of the techniques employed by the hen cuckoo in depositing her egg. This particular photographic success particularly delighted Edgar Chance since he could now confirm that the cuckoo laid its eggs directly into the nest of its host by means of an extending cloaca. Armed with Oliver Pike's film he was now able to disarm his critics, although a few, such as Percy Bunyard remained stubbornly, and vociferously, unconvinced.

Although this photographic work was carried out in 1922, the book, "The Truth about the Cuckoo" containing accounts of the study was not published until 1940. It also included amongst its illustrations stills from an earlier cine film taken by

Mr. Hawkins, and many still shots taken by Miss E.L. Turner.

In order to expand and flourish, ornithology needs men who not only possess an extensive knowledge of birds, but who are also capable of presenting the subject in an attractive and readily assimilated form to the public. William Henry Hudson was just such a person, an excellent communicator who, through his writing, did much to popularise interest in birds, particularly concerning the always thorny subject of bird protection. He was born in 1841 in Quilmes, a small village some ten miles from Buenos Aires, and from his earliest days spent much of his time studying the wildlife of the Pampas. He received very little in the way of formal education mainly because no satisfactory tutors could be found in such a wild area. In his autobiography he recalled later that one was a drunkard, one had to be dismissed because of his evil temper, and another was too lazy to teach him anything. Consequently as he described his early life, "I ran wild in a wild country". He passed his time, happily enough, riding around the countryside watching and noting its birds and flowers, growing up to care passionately about natural history in all its aspects.

Just before he was fifteen, tragedy struck, he contracted typhus and then, whilst in a badly weakened physical condition, developed rheumatic fever. This combination of serious illnesses left him not only weak in body but in a depressed and unstable mental condition. Unable to resume his formerly active life, he read obsessively and started to correspond with Philip Lutley Sclater to whom he also sent many specimens of Argentinian birds and mammals. This proved to be a very useful contact for Hudson when he arrived in England in 1869.

Hudson never returned to South America, but he never really settled or felt completely at home in England. Perhaps, as an emotional outlet, he began to write books about birds, one of the most important of which was "Argentine Ornithology" in co-authorship with Philip Lutley Sclater. Hudson wrote the field notes and Sclater the technical descriptions for this work which was published in two volumes in 1888 and 1889. A good deal of the material in this book was later rewritten by Hudson and published under the title of "The Birds of La Plata" in 1920. He wrote many popular works on British birds such as "Birds in a Village" (1893), "British Birds" (1895), "Birds in London" (1898), "Birds and Man" (1901) and "Hampshire Birds" (1903). Always nostalgic for the golden days and the land of his youth he wrote an autobiography of his earlier years "Far Away and Long Ago" which appeared in 1918, besides travel books such as "The Purple Land that England Lost" (1885) and "Idle Days in Patagonia" (1893).

Throughout his life his writing clearly reflected his highly sensitive and withdrawn character, and his natural melancholy was in no way helped by the precarious financial situation in which he always lived. Indeed but for the efforts of Lord Grey of Fallodon who obtained a civil list pension for him in 1901 of £150 per annum, he would have been unable to continue with his writing.

His greatest contribution to ornithology was undoubtedly his work towards bird protection for which he was described by Philip Gosse as "The man who has done more, probably, than anyone of our times to foster and encourage a love of all living wild things, and above all of our native wild birds". William Hudson was a member of the first committee of "The Society for the Protection of Birds", and remained on the committee until his death. He wrote a great many leaflets for the Society, and was a constant and diligent letter writer to the editor of The Times and other newspapers on all matters germane to the protection of birds.

In 1918 his health began to fail and he died in August 1922. W.H. Hudson is a difficult man to categorise; some would say that he was an ornithologist who was also a writer, others that he was an author first and an ornithologist second, of these assessments the second is probably nearer the truth.

For many who commenced their birdwatching during the late 1920s and early 1930s, the name of Thomas Alfred Coward will be remembered with affection as the author of some delightful books from which they first began to gain their knowledge about birds. Born in 1867 in Manchester Thomas Coward went into the textile trade, a job he hated and from which he retired at the earliest opportunity. He was then free to devote his energies to writing for newspapers, especially "The Manchester Guardian" (as it was then), magazines and, of course, books on birds. He also gave lectures and was amongst the very early broadcasters who gave talks on natural history. He always wrote well and clearly, especially about the birds, mammals and history of his beloved native county of Cheshire, much of this work appeared in "The Vertibrate Fauna of Cheshire and Liverpool Bay" (1910) of which he was co-author with Charles Oldham.

His best known work was undoubtedly "Birds of the British Isles and their Eggs", issued in three volumes between 1920 and 1926. This provided a much needed and authoritative book on British birds, containing good coloured plates at a reasonable price which provided a very wide audience with a useful introduction to the study of birds. The plates were taken from Lord Lilford's "Coloured Figures of the Birds of the British Islands" printed in a much reduced format and supplemented by many black and white photographs as well as some text figures. This work was later abbreviated and simplified by Enid Blyton and was published as one small book in 1936 under the title of "Birds of the Wayside and Woodland". It was an excellent pocket sized guide to British birds in which the coloured plates were still retained, and although for this edition they were reduced to about 1½" x 2", nevertheless the quality of the reproduction remained high.

Amongst the other books written by Thomas Coward were "Bird Haunts and Nature Memories" (1922), "Bird Life at Home and Abroad, with other Nature Observations" (1927), "Bird and other Nature Problems" (1931) and "The Life of Birds" also published in 1931. Always a "field man" Coward had little interest in the niceties of taxonomy and other technicalities, but his lively, accurate and informative books were of the greatest help and encouragement to numerous children and amateur birdwatchers.

He was an active and level-headed conservationist, the secretary of the local RSPB group, and a campaigner against all blood sports. He died following a heart attack in 1933. An interesting reflection on the way in which field ornithology has advanced since Thomas Coward's day is provided by his comment, "I find pipits in the field almost hopeless, and not much better in the hand" — most ornithologists today would consider this to be a remarkable statement from an acknowledged field observer.

Whilst men such as Hudson and Coward were spreading interest in, and knowledge of, birdlife in Britain, work of equal importance was being carried out in the United States by Frank Michler Chapman. Born in West Englewood, New Jersey in June 1864 he was educated at Englewood Academy but instead of going to university decided to take up an appointment with the bank of which his father was the legal advisor. Frank Chapman never considered that the time he spent at the bank was wasted since, as he said, it taught him the twin virtues of orderliness and promptness. He spent much of his precious sparetime in birdwatching, a hobby he found increasingly absorbing, until in 1886 he resigned from the bank and commenced an ornithological career. He began by making a collecting trip to Florida where his mother had bought a cottage; at this time the state was practically unspoilt and he gained a great amount of experience from the abundant birdlife. In 1887 he was fortunate enough to make the acquaintance of George Sennet, an amateur ornithologist, who arranged for Frank Chapman to assist him with his bird collection which was housed at the American Museum in New York.

Whilst working at the Museum he took a part-time job with the Director, Dr. J.A. Allen. This entailed sorting a collection of 8,000 specimens of South American birds, and it was under Dr. Allen's expert tuition that Chapman's career as a professional ornithologist could truly be said to have begun. In 1888 he was offered a position as assistant to Dr. Allen, at the princely salary of $50 a month, thus starting a professional association with the Museum which was to last for 54 years, during which time he was instrumental in raising it to the highest rank amongst the world's scientific institutions.

Much of Chapman's taxonomic research was concentrated on neo-tropical birds, which was to culminate in his two great treatises, "The Distribution of Bird Life in Columbia" (1917) and "The Distribution of Bird Life in Ecuador" (1926). Both these books contained detailed geographical and ecological analyses of the areas covered, and also set out his theories concerning the reasons for discontinuous distribution, which is so often a feature of mountainous regions. In preparation for the Ecuadorian book Frank Chapman spent months travelling and collecting in some very difficult terrain, and critically studied some 13,000 specimens; a very similiar workload was involved in his Columbian book.

He did not neglect the birds of his own country, and his "Handbook of Birds of Eastern North America" first published in 1895 went through numerous editions, whilst a primitive fieldguide "Color Key to North American Birds" published in 1903 was a pioneer of its kind in North America. This book contained 800 drawings by Chester A. Reed. The colouring throughout was weak and only prominent features where colour was important as a means of identification as, for instance, the red head and breast of the red-breasted sapsucker, were tinted, the remainder being in black and white. These illustrations were accompanied by brief written descriptions, measurements and call notes etc. An interesting example of current ornithological thought is provided by an excerpt from Chapman's preface to his book, — "From the scientific point of view there is but one satisfactory way to identify a bird. A specimen of it should be in the hand in order that its form, color, and size may be accurately determined....". The aim of his book was to make this practice no longer necessary for the amateur birdwatcher, and he was always at pains to point out that birds should not be collected indiscriminately.

During his long life Frank Chapman wrote some seventeen books and about 225 articles which appeared in various journals. His more popular books did much to stimulate interest in birds and birdwatching, being written in an easy style and well illustrated by photographs. "Bird Life, a Guide To the Study of Our Common Birds" (1897), "Camps and Cruises of an Ornithologist" (1908), "My Tropical Air Castle, nature studies in Panama" (1929) , "Autobiography of a Bird Lover" (1933) and "Life in an Air Castle" (1938) all give informative and interesting pictures of his life and work. One of his expeditions was to study and collect specimens on Mount Roraima, that strange red sandstone block on the borders of Guyana and Venezuela, which had formed the setting for Conan Doyle's "The Lost World". This plateau contains many rare and interesting birds and mammals, but not the dinosaurs and other prehistoric creatures attributed to it in Conan Doyle's novel.

Frank M. Chapman was appointed Associate Curator of the Joint Department of Birds and Mammals of the American Museum in 1901, a position he held until 1920 when a separate Department of Birds was established, of which he became the Curator, an office he held until his retirement in 1942. He spent his remaining years in Florida, still actively studying birds before returning north to Nantucket in the summer of 1945, he died in New York in November of that year.

Apart from his work in South America, he spent much time at Barro Colorado in the Panama Canal zone. He first visited England in 1905 and again in 1921, this time as the guest of Lord Grey of Fallodon with whom he had become friendly during the latter's posting as British envoy in Washington. He went to the Rothschild Museum at Tring, and visited the Itchen Valley and the New Forest in Hampshire, excursions that he enjoyed so much that he repeated them when he made another journey to England in 1932.

Amongst new developments which he instigated and promoted in the American Museum was the creation of displays of "Habitat Groups" in which the exhibits were placed in diorama. They are quite common in museums today, but when in 1901 John L. Cadwalader, a prominent New York sportsman, saw such a display in the British Museum he was so impressed that he donated a sum of money for one to be made in the American Museum. Frank Chapman was equally enthusiastic and went ahead with plans to make a beach scene, but here he met with considerable opposition from his colleagues, particularly with regard to the painted background, which they considered to be a violation of the museum tradition. Feelings ran so high that Chapman finally had to appeal to Morris K. Jesup, the President of the Museum, to act as arbiter. Mr. Jesup was ill at the time but when eventually he was able to view the diorama from his wheel-chair he was immediately impressed, "That is very beautiful" he said and promply gave his permission for the museum to feature this type of exhibit.

Frank Chapman's inspired direction ensured that during his curatorship the American Museum always remained in the van of the movement which strove to make museums more exciting and informative places, whilst at the same time pursuing professional and forward looking scientific policies. One of the most articulate ornithologists of his time, he was an educator who through his writings and lectures won many converts to the cause of conservation and ornithology. He retained his active mind until the end of his life, as was shown by his continuing acceptance of new ideas and his constant encouragement of the young.

X

The Academics

Between the years 1910 and 1939 public interest in ornithology as a leisure pursuit steadily increased, but within the professional sphere, particularly in the fields of behaviour and taxonomy, no such advances had been made; with the result that Britain lagged far behind Germany, Holland and the United States. Amongst the professional naturalists many still adhered to Professor Newton's belief that almost everything of importance had been discovered about birds. Fortunately, as so often happens, after a period of stagnation major changes were on the way and by the mid thirties the movement towards professionalism in British ornithology was once again starting to accelerate, whilst a new generation of men including Max Nicholson, Bernard Tucker and Wilfred Alexander, ably supported by Harry Witherby, were beginning to bring about radical changes in ornithological thought and attitudes.

Although never a professional in the strict sense of the word, the principal force and catalyst in this rebirth of scientific ornithology was Max Nicholson, who, some sixty years later is still playing an important and active part in ornithology in Britain as, along with Stanley Cramp, one of the instigators and editors of 'The Birds of the Western Palearctic''.

Max Nicholson wrote his first book "Birds in Britain" subtitled "An Account of the State of our Birdlife and a Criticism of Bird Protection" in 1926 whilst still an undergraduate at Oxford. In this he set out, within a historical context, some of his ideas for bird protection, especially with regard to agricultural changes and urban sprawl, many of which remain as valid as when they were originally propounded. Nor, although the emphasis has altered somewhat, are many of his strictures upon egg collecting entirely irrevelant today, even if nowadays the main reason for taking eggs is to endeavour to rear the young of rare birds, such as falcons, to sell for high prices abroad, rather than the former practice of amassing vast numbers of eggs simply for display.

Max Nicholson's greatest achievement was the leading part that he played in the formation of the British Trust for Ornithology — a title he chose himself. The original idea developed from the Oxford Ornithological Society, founded in 1921, and in which he had been a prominent member. He organised the setting up of the Oxford bird census, established a ringing station in Christchurch meadows in 1927, and in 1928, with the support of Harry Witherby, organised a national census of heronries. All three projects or, more precisely, developments of them, the common breeding bird census, bird observatories carrying out ringing projects, and censuses of selected species together with the long running census of rookeries, still form the major branches of the British Trust for Ornithology's activities today. These far-ranging undertakings epitomise the aim of the British Trust for Ornithology, which is to plan and organise ornithological research on behalf of its members, and to provide a trained scientific staff who have the necessary professional expertise to extract, analyse, interpret and coordinate the great mass of data produced by a large body of amateurs.

"Concentration of aim, expert direction and the training of observers" was the blueprint for progress deemed necessary by Max Nicholson and the other co-founders Bernard Tucker, Wilfred Alexander and Guy Charteris.

They fully appreciated just how much more there was and still is to discover about birds, and whilst always insisting upon and emphasising the importance of the interlocking roles of amateur and professional, realised from the beginning that much of the work demanded not only a professional training, but the time and facilities which can only be provided by a permanent body of experts. Thus, over the years, a whole new class of professionals was created, with profound and far reaching effects upon ornithology not only in Britain but in some directions, notably bird census taking, worldwide.

The literature produced by these developments was mainly in the form of papers in scientific journals, although many of the books which have been published since its formation owed a great deal to work carried out by the British Trust for Ornithology.

Because of the pressures of his outside work Max Nicholson's literary output has been restricted, but "The Art of Birdwatching" (1931) sets out in detail his ideas on this subject, many of them extremely advanced and innovative for their time. "Birds and Men" published in 1951 in the New Naturalist series was a scholarly treatise on those species of birds which are most closely associated with man. He was also amongst the pioneers in yet another field, that of the "sound guide" to birdsong. During the course of the last thirty years we have come to accept and indeed take for granted, records or tapes of birdsong accompanied by a book or set of notes, as a most useful tool in identifying birds. The first such guide was however produced as long ago as 1936 when "Songs of Wild Birds" was written by Max Nicholson as a complement to two records by Ludwig Koch, this was followed in 1937 by "More Songs of Wild Birds" together with three records, both sets of books and records being issued as a "boxed set". Given the impediments existing at the time these books and records set an astonishingly high standard, but further development was largely dependent on the introduction of lightweight recording apparatus and upon long playing records and cassettes to replace the heavy discs capable only of playing for about eight minutes on each side.

Following the oubreak of the second world war circumstances forced Max Nicholson to hand over his position as honorary Secretary of the British Trust for Ornithology to James Fisher. As a senior Civil Servant he had been appointed as secretary to the Office of the Lord President of the Council, and in this capacity he was present at some of the most important meetings which took place during the war. Later, from 1952-1966 he was a highly successful Director General of the Nature Conservancy and also an editor of "British Birds". Finally, at an age when most men contemplate taking life somewhat more easily, he became deeply involved in his work as one of the editors of the "Handbook of the Birds of Europe, the Middle East and North Africa", which is probably the most important single bird book to be published in Britain since Witherbys Handbook was completed in 1941.

As a corollary of his lifelong work for conservation both as an amateur and later professionally with the Nature Conservancy Council his book "The Environmental Revolution" published in 1970 is a formidable treatise on the problems of conservation worldwide.

Max Nicholson's part in the founding of the British Trust for Ornithology and his co-editorship of "British Birds" reinforced his connections with another pioneer and driving force in British ornithology, Harry Forbes Witherby, who over a period of fifty years played an increasingly important role in the ornithological world. Indeed it would be true to say that from the end of the first world war until his death in 1943 there was hardly any development in which he was not in some way concerned — frequently as a moving spirit — with the sole aim of raising the standard of ornithology in Britain to a consistently high professional level.

Born in October 1873 at Burley, he grew up surrounded by the beautiful Hampshire countryside from whence came his lifelong love of, and interest in, birds. His first articles appeared in the magazine "Knowledge", and his first book, "Forest Birds, their haunts and habits" was published in 1894. In this small book he describes in some detail many of the common birds of his beloved New Forest. On leaving school he joined the family publishing firm of "Witherby & Co.", later to become H.F. & G. Witherby, where he spent the remainder of his working life. Over the years his connection with publishing was of tremendous value to British ornithology, many important books were to bear the Witherby imprint, including the classic "Handbook" and its predecessor the "Practical Handbook".

Of equal import was the founding in 1907 of the monthly magazine "British Birds", this provided a much needed authoritative source of notes and articles, mainly concerning birds in Britain, in a rather less specialised form than those offered by "The Ibis". With the exception of a short period during the first world war, when he was away on active service, Harry Witherby edited this magazine up until the time of his death.

Once "British Birds" was firmly established Harry Witherby started to plan the publication of an up-to-date, concise and accurate book which he felt was required to provide information for the rapidly growing number of people who were actively interested in birdlife. The first part of "The Practical Handbook of British Birds" was in fact already printed prior to the outbreak of war in 1914, but its issue was suspended. It was eventually published, (in two volumes bound in three parts) between 1919 and 1924, when it was immediately recognised for what it was, a milestone in ornithological literature. Unusually for its time it was the work of a number of specialists, each contributing a piece on each bird, under the strict editorship of Witherby, who himself wrote the description and distribution sections. Nomenclature and keys were the work of Ernst Hartert and the detailed plumage descriptions, which were classics of their kind, were by Anne Jackson, later to become Mrs. Meinertzhagen. F.C.R. Jourdain, Charles Oldham and Norman Ticehurst were the other contributors.

Excellent though the "Practical Handbook" was, Harry Witherby was still not entirely satisfied, and he proceeded to set about planning the publication of a much longer and more comprehensive work which he intended to be fully illustrated in colour throughout. Like its predecessor it was under joint authorship, F.C.R. Jourdain, Norman Ticehurst and Harry Witherby this time with the addition of Bernard Tucker. "The Handbook of British Birds" was issued in five volumes the first appearing in 1938 and the last in 1941. This work has become one of the classic birdbooks, it was undoubtedly the greatest single contribution made to ornithological publishing in Britain between the wars, and stands today as a permanent memorial to its hard-working authors.

Despite the demands made on his time by his business interests and the vast amount of work involved in editing "British Birds" and "The Handbook", Harry Witherby held the position of Honorary Secretary and Treasurer of the British Ornithologists' Club and was also President of the British Ornithologists' Union from 1937 to 1938, whose highest award, the Godman-Salvin medal, he received at the completion of his presidency.

Always ready to assist the progress of ornithology in Britain in any way possible, a fine example of his practical help and generosity occurred in 1933, when due to lack of funds the British Trust for Ornithology looked likely to die in infancy. In order to ensure its survival Harry Witherby sold his large collection of bird skins to the British Museum for £1,400 and gave the entire proceeds to the trust. So all pervading and great was his influence that his death in 1943 was keenly felt in every section of the ornithological establishment in Britain.

Bernard William Tucker who took a large part in producing "The Handbook" was one of the finest professional ornithologists of his day. Although less well known than some of his contemporaries, this comparative anonymity is mainly due to his early death at the age of 49.

He was born in 1901 in Hertfordshire, but spent much of his youth in Somerset. He was educated at Harrow and went up to Magdalen College, Oxford in 1919 where he gained a first class honours degree in Zoology in 1923. His zoological interests were widespread and by no means confined to the study of birds: he did a great deal of research work on reptiles, and from 1924 to 1925 worked as a marine biologist in Naples. For a short while in 1925 he went to Cambridge as a demonstrator in zoology before returning to Oxford in 1926, where as an acknowledged authority on comparative anatomy he acted as tutor, lecturer and demonstrator besides devoting much of his time and expertise to various University committees. Recognition of his outstanding work came in 1946 when he became the first Reader in Ornithology at Oxford, the first such appointment to be created at any British university.

As his friend and colleague, Max Nicholson, quite rightly pointed out in his obituary, much of Bernard Tucker's work was done in close collaboration with others and therefore it was difficult to separate his own particular contribution. Apart from his involvement, as one of the four contributors to "The Handbook"; from 1943 until his death in 1950 he was editor of "British Birds", taking over from Harry Witherby at a time when the production of the magazine was beset by wartime difficulties.

There can be no doubt that had he lived longer he would have made a greater impact not only on the magazine but on the entire British ornithological scene. He played a considerable part in the founding of the British Trust for Ornithology, but of even more importance was his influence upon the direction of the upsurge of interest in ornithology which immediately followed the war. His insistence on accurate and methodical work ensured that this newfound enthusiasm was channelled into fruitful investigations, properly and scientifically organised rather than in a haphazard and random fashion. Despite great suffering he continued to work almost until his death, and was preparing copy for "British Birds" in hospital almost to the day he died. In his memory, and in recognition for all that he did for the British Trust for Ornithology, the Bernard Tucker medal is awarded each year to the candidate who has contributed most to field ornithology in the Trust enquiries.

Another Oxford ornithologist who through his active membership of the Oxford Bird Club played a prominent part in the setting up of the British Trust for Ornithology and many other aspect of British ornithology was Francis Charles Robert Jourdain. Born in March 1865 at Ashbourne-cum-Mapleton in Derbyshire, he was an undergraduate at Magdalen College, Oxford and ordained in 1890. After holding curacies in Suffolk he became the vicar of Clifton-by-Ashbourne, a living he held for twenty years before moving, in 1914, to Appleton near Abingdon in Berkshire, where he remained until his retirement in 1925.

Although he is today best remembered for his participation in Witherby's "Handbook" and its predecessor, his erudition and enthusiasm made Francis Jourdain one of the leading amateur ornithologists of his time. He published very little work under his name alone, but several books such as a treatise on the eggs of European birds, and a book on the birds of Spitsbergen were left incomplete at his death. Much of his writing was in the form of papers and articles for scientific journals such as "The Ibis", "British Birds", the "Oologist's Record" and "The Auk" etc. He spoke and wrote fluent German and contributed several papers to "Journal fur Ornithologie" as well as translating Kleinschmidt's "The Formenkreis Theory".

In 1921 he organised and led the first Oxford expedition to Spitsbergen, a journey he repeated in 1922. He travelled widely in Britain and throughout the Western Palearctic region in furtherance of his great hobby of egg collecting, but in spite of his intense interest in this subject he was no nest plunderer, only taking eggs which because of some peculiarity in their marking he thought might illustrate an important point. Naturally his egg collecting did not endear him to many people, and neither did his sharp tongue. When disagreeing with some statement he was apt to be both combative and argumentative, and not for nothing was he known as "Pastor pugnax". This acerbity was rarely more in evidence than during meetings of the British Ornitholgists Club, especially if Percy Bunyard, another fervid egg collector was present. Records of the early thirties show that attendances at the British Ornithologists Club's suppers were frequently extraordinarily high, an occurrence not unconnected with the fact that the rows between the two were often well worth hearing — Jourdain employing withering sarcasm whilst Bunyard favoured vulgar abuse. A spectator of some of these historic skirmishes recalls that on such occasions it was unwise to sit next to Bunyard at table, since if Jourdain was speaking he would prod his neighbour to ask, in the peculiarly loud voice of the deaf, "What bloody lies is he telling now?". Jourdain however was a man who always made very sure of his facts before committing himself to any statement either written or verbal.

He was a world authority on the breeding biology of the birds of the Western Palearctic region, and on the food and geographic distribution of species. From 1917 to 1919 he took over the editorship of "British Birds" during the period that Harry Witherby was in the navy, and remained on the editorial committee until his death in February 1940.

Although his wide knowledge and all round ability was of the greatest value to ornithology in Britain, he received very little in the way of formal acknowledgement in ornithological circles; nor, despite serving as Vice President of the Union from 1934-1936 was he offered any of the British Ornithological Union's awards. This lack of recognition was undoubtedly partly due to the dislike and suspicion in which egg collectors were held, and also to Jourdain's critical tongue and punacious nature.

At much the same time that the British Trust for Ornithology was struggling to come into existence, a sibling organisation was also being formed in Oxford, the Edward Grey Institute. This centre for ornithological research was first proposed as a memorial to Lord Grey of Fallodon, a prominent politician who later became Chancellor of Oxford University. A life long student and lover of birds he did much to encourage the proposal that there should be a national ornithological research centre based in Oxford. Although he died in 1933 the Edward Grey Institute was not officially set up until 1938, with Wilfred Alexander in charge of the library. This now bears his name, and houses one of the largest collections of 20th century bird books and manuscripts in Europe.

In any account of the foundation of the British Trust for Ornithology and the setting up of The Edward Grey Institute the name of Wilfred Backhouse Alexander must be included as one of notable importance: not only for his constant support and encouragement of these schemes, but because in his position as the first fulltime research officer in the Oxford University Research Programme into economic ornithology he was officially funded by the Ministry of Agriculture, and thus was the first of a new breed of professional ornithologists whose opinions carried considerable weight far beyond academic and ornithological circles.

Born in Croydon in 1885, he spent most of his youth in Tunbridge Wells before going up to Cambridge to read natural sciences, specialising in botany. In 1912 he was offered a museum appointment in Western Australia where he remained for some fifteen years. His most notable achievement during this time was the introduction of the Cactoblastis insect from Argentina, using it as biological agent against the prickly pear which had earlier been introduced into Australia and become a serious pest. Returning to England in 1926 he commenced work upon his book "Birds of the Ocean" which was published in 1928 and being an identification guide was a prototype for every fieldguide which has followed. Surprisingly, its concept was not taken up in Britain until some time later, although Roger Tory Peterson working quite independently in the United States developed and improved the same idea in his field guide published in 1934.

Wilfred Alexander spent a year as superintendent of the Marine Biological Association's Survey of the Tees estuary. Upon his return to Oxford in 1931 he set to work on the surveys and bird census which had been started by V.C. Wynne-Edwards, an eminent biologist who later became Regius Professor of Natural History at Aberdeen University from 1946 to 1974. Whilst all credit must be given to Alexander for his scientific knowledge, particular mention must be made of the skilful manner in which he persuaded and encouraged his associates to carry out those studies for which he was ultimately responsible. His diplomatic talents were of the utmost importance in census enquiries, such as that on the herons, which depended entirely upon the cooperation of numerous observers and local societies. As succeeding generations of census organisers have all discovered, a successful enquiry requires a great deal of tact and not a little gentle prodding.

He is probably best remembered today for his part in the building up of the national ornithological library at Oxford which bears his name. He was its first librarian and his generous contributions from his own extensive library formed the original basis of the collection. Although highly esteemed by many of his peers, Wilfred Alexander has never received full recognition for the important work he did towards inducing a more scientific approach to ornithology. He died in December 1965 at the age of 80.

Wilfred Alexander was one of three brothers, all of whom were quite exceptional ornithologists and each of whom made a significant contribution to their own respective fields. Christopher Alexander was born in 1887 and soon evinced an early interest in natural history, particularly in birds and lepidoptera. Together with his brother Horace, he pioneered a method of mapping territories and counting breeding birds around their home in Tunbridge Wells which, in a modified form, is still in use today. He trained at Wye Agricultural College and in June 1911 went to Rome to work for the International Institute of Agriculture, later to become the Food and Agriculture Organisation of the United Nations. During his time in Italy he continued to study the local wildlife and contributed an important paper to "British Birds" on "The Zonal Distribution of the Birds in Latinum". He was one of the first to appreciate that in mountain areas different bird species occupy different altitudinal "layers". Christopher Alexander would have undoubtedly given much to ornithology, but in 1914, despite the fact that the Alexander family were staunch Quakers, he felt the circumstances were such that he should break with the tenets of his family's faith and fight. He was killed at Passchendaele in 1917.

Horace, the youngest of the Alexander brothers, was born in

April 1889, and he too, like his brothers, took a passionate interest in birds from his early childhood; like them he always adopted a strictly scientific approach, keeping faithful records of all those he saw and heard. His early field work was carried out mainly around Tunbridge Wells, Dungeness and the Romney Marshes, but later when his work took him to the West Midlands he continued to keep meticulous notes and from these observations was able to supply much valuable information for the "Handbook of British Birds", especially on song periods and identification.

The twelve years which Horace Alexander spent in India greatly widened his ornithological experiences, especially concerning the leaf warblers, a group of birds which particularly interested him. Much of his research was incorporated in "Some Notes on Asian Leaf Warblers (Genus Phylloscopus)" a small book published privately in 1969. This forms a most valuable addition to the literature concerning a family of birds whose field identification has always presented peculiar difficulties. Horace Alexander's "Seventy Years of Bird Watching" (1974) gives a fascinating picture of the character and attitudes of various ornithologists as observed by someone who knew most of them well, besides providing a history of ornithology in Britain since before the first World War.

Despite their comparatively small literary output, the Alexander brothers through their inventive and forward looking outlook and opinions played a significant part in the development of ornithology over a period of some seventy years.

One of the foremost scientific ornithologists of the 1950s and 1960s was David Lambert Lack. Born in 1910 he spent most of his childhood in London but was sent away to be educated at Greshams School at Holt in Norfolk. Greshams was very much in the forefront of those enlightened schools which encouraged their pupils to take an interest in activities which lay outside the normal scholastic curriculum and team games. For David Lack places such as Cley, Salthouse and Blakeney, all within easy cycling distance of Holt, offered superb opportunities for watching birds. In 1929 he went up to Magdalene College, Cambridge to read zoology, and where as a choral exhibitioner he was able to indulge in his lifelong love of music and poetry. Kelling Heath provided him with the opportunity to continue with his studies on the nightjar, his favourite bird at that time, and at the age of 22 he published a major paper on its breeding biology in "The Ibis" as well as submitting additional material to "British Birds" which proved conclusively that these birds were indeed "double brooded".

As had been mentioned earlier he was amongst the first of a new generation of serious ornithologists to appreciate the pioneer work of Edmund Selous; besides bringing to the attention of a much wider public the writings of J.P. Burkitt an Ulsterman who had carried out a great deal of original research on the robin. He was not afraid to criticise, and having read Eliot Howard's theories on territory and bird behaviour he produced, after some reflection, some stringent comments on Howard's ideas in an article published in "British Birds".

David Lack took a prominent part in the activities of the Cambridge Bird Club, becoming its secretary-editor and later its president. The Club at that period contained a positive galaxy of future ornithological stars in the shape of Peter Scott, Tom Harrisson, Con Benson, Myles North, Bryan Roberts, and from Australia, Dom Serventy.

As, at this time, there were no opportunities for a fulltime salaried ornithologist, the only professional zoological field open to him was to specialise as an entomologist; and it was in this capacity that he was included as a member of the Oxford-Cambridge expedition to St. Kilda in 1931. This trip was followed by further expeditions to Bear Island and Scoresby Sound in Greenland. During these years some of his most valuable work was concerned with the changes, brought about by afforestation, in the fauna of the Brecklands. Such studies were symptomatic of the growing realisation in the 1930s of the importance of combining the various disciplines of botany and entomology with ornithological studies in order to gain a properly balanced picture.

After coming down from Cambridge in 1933, a chance meeting with Sir Julian Huxley resulted in David Lack taking up a teaching post at Dartington Hall School in Devon. In the first instance this was intended to be simply a stopgap engagement while he waited for an appointment at the British Museum (Nat. Hist.) which was expected to become vacant in the near future. However so enjoyable did he find his work at Dartington Hall that when the British Museum post was offered, he declined it.

Feeling that some experience of tropical conditions would be of benefit to him Julian Huxley suggested to David Lack he should get in contact with Reg Moreau who was then working at Amani, and it was arranged that he should spend six weeks of his summer holidays with the Moreaus in Tanzania. During a visit to the United States in 1935 he met J. Grinnell, an ecologist who was engaged in some highly important behavioural studies: during this meeting ideas for further investigations were discussed which indirectly were to have considerable repercussions on his future work. In 1937 he decided that before he could continue with his detailed study of the robin (commenced some years previously) it was necessary that for purposes of comparison he should first examine the territorial behaviour in another group of birds. Deciding that Darwin's finches would be the family which would best provide him with the material he needed for this particular research he accordingly travelled to the Galapagos Islands in the autumn of 1938. The final stages of this work were interupted by the outbreak of war in 1939, and "Darwin's Finches" was not published until 1947. This book, originally conceived as being suitable to demonstrate evolutionary principles to sixth form pupils, ended up as a highly specialist scientific treatise in which he proved that certain taxonomic problems of speciation can only be satisfactorily decided by ecological studies of the birds in the field.

The war opened up for David Lack yet another line of

research. In 1940 he was approached by the Central Register for Scientific Workers with the offer of a place in an operational research group involved in resolving problems arising from the use of the then top secret radar.

He volunteered for a spell of duty in the Orkneys and from March to August 1941 had what he later described as "a wonderful holiday", since it gave him an excellent opportunity to study seabirds during his off duty hours. As a result of his investigations he was able to identify radar echoes, popularly known at the time as "angels", as being in fact caused by birds: and by means of this wonderful new device he was able to study and interpret the movements of birds, particularly the night migrants, with a hitherto impossible degree of accuracy

He was not only able to advise on the probable cause of the echoes, but was also to use that data to revolutionise the understanding of the numbers and routes taken by migrant birds in Britain After the war David Lack was able to carry out a special study (using a powerful military radar) to prove that swifts, who in their evening flights climb up far beyond the limits of human sight, spend all the hours of darkness wheeling high above the earth.

In the midst of his wartime researches he at last managed to finish "The Life of the Robin", "to relieve" as he put it "the dreariness of boarding house evenings and Sundays". Strangely the first publisher he approached refused the manuscript, but Harry Witherby immediately appreciated its importance and published it. This book, a classic behavioural study of a familiar garden bird was not only very popular with a wide public but also brought his name to the fore in ornithological circles.

During the closing stages of the war he was asked by Bernard Tucker if he would consider taking over the directorship of the Edward Grey Institute after Wilfred Alexander's retirement. Initially he declined this offer but later decided to accept it and so, at the age of thirty-five became a full time ornithologist. Under his leadership the Institute became a recognised force throughout the ornithological world. From the start he felt that the Institute should take a definite and independent line of its own, and since other institutions were already investigating migration and behaviour in some depth, decided that its energies should be mainly concentrated on population studies of birds.

Although perhaps not quite as widely popular as "The Life of the Robin", "Swifts in a Tower" published in 1956 attracted a wide readership. It contained a delightful account of the swifts nesting in the roof of the Zoology Museum in Oxford and remains a valuable species monograph. David Lack wrote a number of other important books including "The Natural Regulation of Animal Numbers" (1954), a major work not solely confined to birds, "Population Studies of Birds" (1966), and "Ecological Isolation in Birds" (1971). These three works represented his greatest contribution to the knowledge of the ecology of birds.

His last book, a valuable treatise on zoogeography, "Island Biology, Illustrated by the Land Birds of Jamaica" (1976) was written in the knowledge that he only had a short time to complete it before he died. It was left to his wife, Elizabeth, and James Monk, working from his notes in order to complete the chapters he had been unable to finish, and see the book through to publication. The theories he propounded in this book did not always agree with those of other workers in this field. It is very much to be regretted that he was never able to expound his ideas further.

Although David Lack had been appointed to the Scientific Advisory Committee of the newly founded British Trust for Ornithology during his days at Dartington Hall, perhaps because of his dedication to the work in hand, he was not much of a "committee man" and took little part in the affairs of the British Ornithologists Union or the British Trust for Ornithology. A shy and rather diffident man, his single minded concentration made him seem withdrawn and remote to some who met him, but many found him to be most sympathetic and helpful. His basic kindliness was never more in evidence than during his "Student Congresses" held annually in Oxford, where budding ecologists were able to freely discuss their work and future projects, often being encouraged by David Lack to undertake lines of study which had previously been neglected.

Despite his total absorption in this work he could on occasion display quite a sense of fun, as shown in his pseudoscientific treatise on "Hardy's swift" (Apus durus), named after Alastair Hardy, later to become Professor of Zoology at Oxford. This description was published in "Bird Notes" (as the journal of the Royal Society for the Protection of Birds was known at the time) and related in great detail how this mythical creature never came to earth, even hatching its young in the air. This article was so convincingly written that many readers wrote in to the BBC to ask for further information in one of its natural history programmes.

Through his attitudes and writing David Lack did much to transform ornithology from a haphazard study into a scientific discipline; but for his untimely death in 1973 he would undoubtedly have played an even greater part in making it the highly professional study that it has become today.

Reginald Ernest Moreau could truly be described as something of a "late starter" in the ornithological world: in his youth he showed little or no interest in birds, but as a young man he became an ardent amateur ornithologist, later to become a full-time museum worker who was much respected particularly for his notable contribution to African ornithology.

Born at Kingston in 1897, he entered the Civil Service in 1914, because of his chronic ill health he was considered to be unfit for active service and worked in the army audit office throughout the war. After the war, since his doctors felt it was imperative that he should live in a warm dry climate, he obtained a posting to Cairo. Here he met Michael Nicoll, an authority on Egyptian birds, and whose knowledge and enthusiasm first awakened Reg Moreau's interest in ornithology. This was further reinforced by his friendship with C.B. Williams, who, as a trained entomologist, constantly impressed upon Moreau the importance of a correct and scientific method of approach to all matters concerning natural history.

In 1927 he managed to obtain a position as secretary and librarian to the Biological Station at Amani in Tanzania, where by a fortunate coincidence C.B. Williams was also a member of the staff. Although initially relatively untrained and inexperienced, by 1930 he was considered to be sufficiently proficient to contribute two specialised chapters, on migration and the birds of ancient Egypt, to "Nicoll's Birds of Egypt". He subsequently produced a steady stream of papers on the little researched birds and tropical ecology of Africa. He also studied epiphytic orchids and african violets. Geographically the area around Amani was particularly suitable for a research station in that it had within a fifty mile radius, hills providing an altitudinal range of some 8,000 ft. with an an annual rainfall varying from 10″ to 100″.

Despite its ideal situation some formidable difficulties faced Reg Moreau and his collegues. Formerly a part of the German Empire the few books that there were on the flora and fauna of the region were out of date and written in German. Quite undaunted Reg Moreau set out to discover for himself which species of birds inhabited the area; in this undertaking he was encouraged and helped by Admiral Lynes, a specialist in African birds, and through whose influence the Natural History Museum in London assisted Moreau in identifying many previously unrecognised birds. An important element in his success was the careful and reliable work done by his African assistant, Salimu Asumani, who not only obtained many of the specimens but whose notes and observations formed a valuable source of information.

Reg Moreau also received much help and advice from Dr. James Chapin of the American Museum of Natural History, who was at that time engaged in the research for his great book on "The Birds of the Belgian Congo". Another visitor to Amani was Dr. W.H. Thorpe, then primarily an entomologist engaged in the study of coccids, but later to become better known as a world authority on bird song and, today, Professor of Animal Ethology at Cambridge. David Lack's visit in 1934 not only provided him with some much needed assistance, but also gave a fresh impetus to his researches.

At the time that Reg Moreau was working in Africa, the study of ecology, especially the ecology of tropical forest and savannah was in its infancy. The papers which he contributed to "The Journal of Ecology" and "The Journal of Animal Ecology" were pioneer works, and his appreciation of the significance of the different zones of animal and plant life at different altitudes and his recognition of the discontinuous distribution of bird species were important and original discoveries. His time at Amani, with its resident professional staff and regular procession of distinguished visitors marked, as he said himself in a biographical note, "the beginnings of the education I never had".

During his time in Tanzania he and his family suffered greatly from the climate and tropical diseases, especially during the second world war when he was unable to take any leave. In 1946 he developed a severe eye condition which forced him to return to England for treatment. His eyes improved but it was clear that he must take early retirement and remain in Britain. He had hoped to be able to take up a position at the sub-department of Animal Behaviour at Cambridge, but delays and difficulties arose, and when, thanks to David Lack, he was offered a place at the Edward Grey Institute he was delighted to accept. Here he flourished academically, not least as editor of "The Ibis" where over a period of thirteen years he greatly improved the standard of that journal: whilst through a judicious blend of tact and efficiency he managed to retain the esteem and friendship of authors whose occasionally pompous prose he never hesitated to prune ruthlessly if necessary.

During his time at the Edward Grey Institute his theories on bird migration gradually evolved and took shape, the culmination of which was his great book, "The Palaearctic-African Bird Migration Systems" published in 1972, two years before his death. He was amongst the first to fully appreciate that migrants returning to Africa fly over the Sahara Desert on a broad front and not channelled into narrow bands down the Nile valley and around the coast of West Africa as had been previously assumed.

Perhaps his greatest contribution to the ornithology of Africa was "The Bird Faunas of Africa and Its Islands" published in 1966, a book which clearly displayed the breadth of his knowledge and understanding of African birds. He then embarked, in collaboration with Mrs. B.P. Hall, on another major pioneer work, "An Atlas of Speciation in African Passerine Birds" published in 1970. This was an important taxonomic study as well as a distributional atlas, which was the first to cover such a vast area. Although by this time Reg Moreau's health was failing, he was crippled by arthritis and beset by various other ailments, his influence permeates the book, although this in no way underestimates the contribution made by Pat Hall, who is both a field observer and a museum ornithologist of the first rank.

The importance of his work, both in Africa and England, is all the more remarkable when set against the background of his complete lack of formal scientific training, and the serious gaps in his education caused by his ill-health and financially deprived circumtances of his childhood. In 1960 he was elected as President of the British Ornithologists Union, and was awarded the Godman-Salvin medal in 1966.

His kindness and humility were legendary and he made friends wherever he went. Despite his health he remained enthusiastic and full of plans for future projects and articles, taking a continual delight in life until his death which occurred the day after his 73rd birthday in May 1970.

During the 1930s the work of such men as Ernst Mayr and Charles Vaurie removed the focal point for the study of taxonomy from Europe to the United States. Charles Vaurie was born at Beaulieu sur Dordogne in France in 1906. As a young man he emigrated to America and settled in New York where he completed his high school education. He studied at

New York before going on to the Dental School of the University of Pennsylvania and qualified in 1928. As a keen amateur bird artist he was introduced, in the early 1940s, to James Chapin, a leading worker at the American Museum of Natural History, and as a result met a number of professional ornithologists who helped to fire his latent enthusiasm for taxonomy. This was especially true of Dr. Ernst Mayr, one of the great taxonomists of the 20th century, under whose guidance Vaurie learned how best to employ his particular talent for meticulousness and care so necessary for this subject.

In 1942 he joined the Ornithology Department of the Museum as a voluntary worker, and in 1946 became an amateur research assistant. A joint work with Ernst Mayr published in 1948. "Evolution of the Family Dicruridae" established him as a taxonomist of great potential.

Gradually he devoted more and more of his time to ornithological work until in 1956 he finally joined the staff of the American Museum as an assistant curator, becoming its curator in 1967. He contributed a great many papers to scientific journals on numerous families of birds, besides a major work, "Birds of the Palearctic Fauna" published in 2 volumes in 1959 and 1965. In this classic work he steered a clever course between "splitting" families into too many subspecies or "lumping" together several species into one. This book codified and concentrated his theories formulated originally in a series of 53 issues of the "American Museum Novitates" which amounted in all to a staggering 958 pages.

An excellent linguist, he spent some time in 1957 in field studies in Cuba and visited Guadeloupe in 1961. He visited Europe frequently and made several trips to the USSR where he became the best known American ornithologist of his day. In 1972 he published in England a substantial book entitled "Tibet and its Birds". Although, due to political circumstances, he was never able to visit Tibet, this is an interesting and scholarly work concerning a region about which comparatively little is known, ornithologically or otherwise. It illustrates very clearly Vaurie's meticulous approach to his work, his deep and thorough research extended to producing a complete gazeteer of the country, a history of all the ornithological explorations, as well as species accounts and surveys of the faunal regions.

A hospitable man and an amusing raconteur, Charles Vaurie was a noted and dedicated worker at the Museum, although not prepared to undertake any more administrative work than was absolutely necessary. In all he published more than 150 papers on birds, and was contemplating a work on the birds of Arabia, but his death, in 1975, prevented anything more than a general outline being prepared.

During the course of an extremely long and active professional career, Alexander Wetmore was to become the doyen of 20th century American ornithology. He was born in June 1886 in North Freedom, Wisconsin, the son of a country doctor, and from a very early age displayed an absorbing interest in birds. For a short while he worked on the Santa Fé railroad, and then, at the age of twenty, got his first museum job at the University of Kansas, the University from which he obtained his degree. In 1910 he went to work for the Department of Agriculture on whose behalf he commenced some specialised studies on the food intake of birds. He was later sent to Washington where he made the acquaintance of many famous contemporary ornithologists, men such as C. Hart Merriam, Leonard Stejneger and Robert Ridgway. In 1911 he was sent to Puerto Rico and in 1920 to South America in order to study the migration of species from northern to southern America. He led an expedition to the mid-Pacific in 1923 where he was in time to study species of birds which have since become extinct.

In 1925 Wetmore was appointed as Assistant Secretary of the Smithsonian, a post he held for almost twenty years. This was a great honour; he was only the second ornithologist to be assigned to this position, after Spencer Fullerton Baird. He devoted a great deal of his time to the American Ornithologists Union and was its President from 1926-1929 and Honorary President from 1975-1978.

He worked for many years in close association with the National Geographic Society, and was the principal author-editor of three of their popular birds books, "Book of Birds" (1937) "Song and Garden Birds of North America" (1964) and, "Water, Prey and Game Birds of North America" (1965). Amongst his more scientific works were "Observations on the Birds of Argentina, Paraguay, Uruguay and Chile" (1926), "The Migration of Birds" (1927) and in 1931 "Birds of Haiti and the Dominican Republic".

Alexander Wetmore's "Systematic Classification for the Birds of the World" was widely accepted and generally referred to as "The Wetmore Order", and from the 1950s until the mid 1970s it formed the framework around which most bird books and checklists were written.

His last great work, left incomplete at the time of his death, was "The Birds of the Republic of Panama" published in three volumes in 1965, 1968 and 1972. He was unable to complete the fourth and final volume because of failing health and his associates brought the work to publication in 1984. Far from being a dry treatise written in the comfort of a museum armchair, this book is based solidly on personal field experience, the result of many lengthy visits to Panama over a considerable number of years.

Although he left the Smithsonian officially at the age of 66, Alexander Wetmore maintained his office there for another twenty-five years after his formal retirement. His output of papers and articles was enormous. In total he described some 189 new species and subspecies; and is unique amongst ornithologists in that a glacier in Antarctica was named after him.

As a taxonomist, an avian paleontologist, a specialist in bird migration, distribution and classification, Alexander Wetmore made his mark in many branches of ornithology. Such widely varied interests would be quite impossible in the highly specialised environment of professional ornithology today. His death in December 1978 in many ways marked the end of an era.

Amongst the many outstanding but scientifically untrained

enthusiasts who have made an appreciable contribution to their country's ornithological literature, it is important that the work of Gregory Macalister Mathews should not be overlooked. Born at Bramble in New South Wales in September 1876, he was educated at Kings School, Parramatta, where he soon developed a wide interest in natural history subjects. As a young man he worked for a time on a cattle station and this was followed by a spell of prospecting for gold in Queensland. In 1902 he left Australia for England, settling at first near Southampton before moving to Watford in order to be within easier reach of the museums at London and Tring. Whilst still at Southampton he started to outline the plan of a book on the birds of Australia. Very little had been published, (apart from E.P. Ramsey's "Tabular List of Australian Birds" in 1888), since the works of John Gould. His close association with Richard Bowdler Sharpe at this time (with whom he worked on the "Reference List to the Birds of Australia"), was reflected in their shared preference for the already obsolescent binominal nomenclatural system: Many of the Genera he named in this period were therefore only temporary. Later he became more orthodox in his attitude to classification.

In 1910 he published the first part of his "Birds of Australia" which had the distinction of being the last of the major books to be illustrated by hand coloured plates, and which, amongst others, included such illustrious names as Keulemans, Gronvold and Lodge.

This great book was published in twelve volumes, in a limited edition of 225 copies issued in 75 parts between the years 1910 and 1928: volumes being updated by means of "supplements" or appendices incorporated in the main body of the work. This included a checklist of Australian birds and "A Bibliography of the Birds of Australia". A later book, published in 1928, "The Birds of Norfolk and Lord Howe Islands, and the Australasian South Polar Quadrant" was, in effect, a supplement to this great work, to which he added a further and final supplement in 1936.

In 1921 Gregory Mathews embarked on yet another project,

this time in conjuction with Tom Iredale, who at one time acted as his private secretary. Sadly, only one volume of this book entitled "The Manual of Australian Birds" was ever published. At the suggestion of the British Ornithologists Union he did however compile the "Systema Avium Australasinarum" which was published under their imprint in two volumes in 1927 and 1930.

In later life he became deeply interested in the petrel family and began work on a book concerning this group which never got beyond the manuscript stage; but a lengthy paper written by him on this family containing a review of the genera and species and accompanied by an identification key was published in the "Bulletin of the B.O.C." in 1948.

Besides building up a large and comprehensive collection of Australian bird skins, (which he later sold to Lord Rothschild), Gregory Matthews assembled a magnificent library, which he presented to the Australian nation in 1939, returning there in 1940 in order to arrange for its installation in the Commonwealth National Library in Canberra. He remained there throughout the war, returning to England in 1945 where he stayed until his death in 1949.

The particularly large variety of seabirds which occur in New Zealand, has naturally attracted a great deal of attention from ornithologists, prominent amongst whom was Lancelot Eric Richdale. Born January 1900 at Marton, North Island, he was educated at Wanganui before going on to Hawkesbury Agricultural College in New South Wales. Returning to New Zealand he taught at a number of schools before taking up an appointment with the Otago Educational Board at Dunedin. Later he attended the University at Otago and received his B.A. in 1934. With no formal zoological training he was given the responsibility of planning nature study courses in the schools besides often teaching the subject to teachers and pupils alike.

In 1936 he began to concentrate his own studies on the seabirds of the Otago Peninsula, the prime object of his research being the yellow-eyed penguin, this was succeeded in the following year by a study of the royal albatross. In all his work on the penguins covered some eighteen seasons and that on the sooty shearwater seventeen. Richdale wrote a major three volume work on the yellow-eyed penguin, the manuscript of which was sent to England during the war but, for various reasons, was never published. His penguin studies were however finally brought together in two books, "Sexual Behaviour in Penguins" and "A Population Study of Penguins" published in 1951 and 1957 respectively.

In 1940 he embarked on an important programme of research on Whero Island with particular reference to the petrels which bred there. The results of this work appeared in a number of scientific periodicals such as "The Auk", "The Ibis", "Proceedings of the Zoological Society of London", "Condor", "Wilson Bulletin" and "Bird Banding". Other papers were published in journals such as "The Emu" and the "Transactions of the Royal Society of New Zealand", and he also, between 1949 and 1952, published a series of ornithological monographs.

As a Fulbright Fellow he visited Cornell university in 1950-1951 whilst a grant from the Nuffield Foundation enabled him to work at the Edward Grey Institute from 1953 to 1955 under David Lack, from whom he received a great deal of help and encouragement.

Throughout his working life he experienced considerable difficulties with both the Department of Internal Affairs and some of his scientific colleagues, probably due to his solitary and reclusive nature — whilst carrying out his field studies he actively discouraged all visitors as he felt they were likely to cause a disturbance. Today he seems to have been somewhat forgotten in his own country and his work is better known and appreciated abroad.

However by any standards Richdale was an outstanding researcher, capable of interpreting his information with great accuracy and detail. Today this mass of data would present few problems, but forced as he was to carry out all his statistical work entirely by himself without the help of a computer, the results he produced were a truly remarkable achievement. A quiet and modest man, he never considered himself to be a professional in the strict sense of the term, but no other word can adequately describe either his method or his results.

Following his retirement Lancelot Richdale returned to London to write up his petrel studies at the Zoological Society under the guidance of Dr. Harrison Matthews. Sadly, around this time he contracted Parkinsons Disease, and was unable to carry out more extensive fieldwork. He died in December 1983.

XI

The Post War Period

The phenomenal growth of interest in ornithology which followed upon the ending of the second world war would hardly have been possible but for two main factors. One was the emergence of a number of talented authors, illustrators and broadcasters, who through the medium of radio and television were able to reach a far wider public than ever before; the other being that this eruption had, to a large extent, been made possible through the work of such men as Harry Witherby, Max Nicholson and Bernard Tucker, who, by creating such institutions as the British Trust for Ornithology and the Edward Grey Institute, had provided a solid scientific foundation upon which much of their work was based.

Immediately after the war the natural history publishing scene was dominated by the launching of "The New Naturalist Library". From its inception this was specifically designed to cover virtually every facet of natural history, botanical, geological, entomological, ornithological and mammalian, with every book having a specialist author for each particular subject. The series, commenced in 1945 had by the end of 1984 reached a total of 69 volumes in the main sequence, with a further 22 special "monographs". So far some twenty books have been devoted to birds, some of them classics of their kind, amongst them "The Fulmár" by James Fisher (1952), "The Wren" by E.A. Armstrong (1955) and "The Hawfinch" by Guy Mountfort (1957). The board of editors for the first twenty-two years consisted of James Fisher, John Gilmour, L. Dudley Stamp and Sir Julian Huxley with Eric Hosking as photographic editor.

Through his work as an author and broadcaster, James

Maxwell McConnell Fisher has, with the possible exception of Sir Peter Scott and Sir David Attenborough, done more than anyone to arouse public interest in natural history. Born in September 1912 he was the son of the headmaster of Oundle School. He won a scholarship to Eton and from there went up to Magdalen College, Oxford, to read medicine, but after taking part in the expedition to Spitsbergen he switched to zoology. After coming down from Oxford he became, for a short time, a schoolmaster, a profession not much to his liking. It was Sir Julian Huxley, who had formed a high opinion of James Fisher's abilities, who really launched him on his ornithological career, being largely responsible for his appointment to the position of Assistant Curator of the Zoological Society of London in 1936.

James Fisher's books covered a tremendously wide range of ornithological subjects ranging from field identification, species monographs, bird watching technique and the avian world in general. He wrote two books concerning his travels in search of birds, "A Thousand Geese" (1953), written with Peter Scott, and "Wild America" (1956) with Roger Tory Peterson as co-author. Perhaps the book by which he will be most remembered was his classic monograph on "The Fulmar" (1952). All birds fascinated him, but seabirds most of all, and another important work was "Seabirds" (1954) which he wrote in partnership with Ronald Lockley. Besides his books he wrote numerous papers on matters as diverse as palaeontology, and the gannet colonies of Shetland: invariably his excellent command of words and erudition made the work not only valuable for its scientific content but also immensely readable.

The main keynote of much of his writing lay perhaps in his strong sense of history, and nowhere is this better illustrated than in "The Shell Bird Book" published in 1966. This rather oddly named work (commissioned by the petroleum company of that name) is an absolute model of its kind, encompassing the composition of the British avifauna, an outline of its history and the place of birds in its literature and music, as well as a review of books on birds, details of societies and conservation besides details of "bird gardening". This book epitomises James Fisher's whole approach to ornithology; a wide historical viewpoint, combined with an impressive thoroughness of research, great breadth of knowledge, and the ability to present information clearly in good plain English untarnished by unnecessary pseudo-scientific jargon.

Despite his academic background he did not take any appreciable part in University or Institutional affairs once he had left the Edward Grey Institute. However the importance of his work was recognised in many quarters, and he was awarded the Union Medal of the British Ornithologists Union, the Tucker Medal of the British Trust for Ornithology as well as the Arthur Allen Medal from the American Ornithologists Union, a rare honour for a Briton. His valuable work for conservation resulted in the R.S.P.B. presenting him with their gold medal in 1961.

His death, as the result of a car accident in 1970, was a tremendous loss to ornithology. As an author he was at the peak of his powers, full of plans for possible future books, amongst those which he was contemplating writing at this time, was a fieldguide to the seabirds of the world, in collaboration with his great friend Roger Tory Peterson, and another on fossil birds which he proposed to call "Early Birds — a Survey of the Fossil Record".

Peter Markham Scott, better known as Sir Peter Scott, is famous worldwide as an author, artist and popular communicator, who has also undertaken a vast amount of serious work both as a pioneer of wildfowl research and a conservationist. Ornithologically speaking he could be said to have been born under a lucky star, starting at his christening when his godfather, Sir James Barry, made him a Life Fellow of the Zoological Society of London. Later, at his public school, he was equally fortunate in that his headmaster was Kenneth Fisher, James Fisher's father, himself a keen ornithologist. During his time at Oundle he was able to watch the great flocks of geese and ducks which wintered on the flooded meadows along the River Nene, and which probably kindled his life long absorption with all kinds of waterfowl, especially geese: an interest he was able to continue when he went up to Trinity College, Cambridge in 1927, although at that time primarily as a wildfowler rather than as an observer.

As special correspondent of "The Field" magazine he travelled to Hungary in 1936 in quest of the wintering wildfowl, and was able to bring back some birds with him from his expedition which he kept at his lighthouse home on the shores of the Wash. During the thirties much of his time was devoted to painting and sailing, but he also wrote and illustrated two books, "Morning flight" (1936) and "Wild Chorus" (1938), which recounted his experiences in wildfowling, his journeys in search of wildfowl and some of his sailing adventures. The outbreak of war in 1939 put a stop to any ornithological work, his wildfowl collection was dispersed amongst his friends whilst he served with distinction in the Royal Navy.

By the end of the war his thoughts had already turned to the possibility of creating a collection of waterfowl which would provide the opportunity both for scientific research and for public enjoyment and education. Various areas were considered, amongst which were the Kent Estuary in Cumbria, High Halstow in Kent and Amberley Wildbrooks in Sussex. Then he received a letter from Howard Davis, an experienced field observer who farmed near Bristol, who suggested that Peter Scott should look at a site in Gloucestershire situated on the banks of the Severn river near the village of Slimbridge. As he later recounted in his autobiography, when the two men met to examine this place they saw wintering flocks of white fronted geese which ran into thousands, besides smaller numbers of bean, greylag, brent and barnacle geese as well as one or two individuals of the extremely rare lesser whitefronted geese. Such a wonderful variety clinched Peter Scott's decision to set up the Wildfowl Trust (as it came to be called) at Slimbridge, and with the active encouragement of the owners of the land, it was to become one of the most important post-war ornithological institutions. Over the years it has grown both physically, by the setting up of "branches", and scientifically with the establishment of a highly qualified organisation to study all aspects of wildfowl biology, with particular attention to conservation and population numbers.

For nearly forty years, in conjunction with the British Trust for Ornithology, the Trust has carried out a countrywide wildfowl count on our inland waters and estuaries. It also does a great deal of research into their diseases and to improving the techniques of keeping and breeding wildfowl in captivity, with the aim of reintroducing them back into the wild. Good examples of the Trust's conservation work is the reintroduction of a large number of Hawaiian geese (NeNe) to their native home, and the white headed ducks to Hungary.

His work as founder and director of the Wildfowl Trust together with that undertaken on behalf of the World Wildlife Fund (of which he is a founder) has naturally taken up a good deal of Peter Scott's time. He did however write a monograph of the swans in 1972, and in 1957 wrote and illustrated "A Coloured Key to the Wildfowl of the World" which succeeded an earlier uncoloured edition published in 1950. An important work in whose production he was also involved, as an illustrator, were the four volumes of "The Waterfowl of the World" written by Jean Delacour and published between 1954 and 1964.

Peter Scott made two expeditions, one to Iceland and the other to the Perry River in Northern Canada, both undertaken with the aim of collecting and studying geese. These journeys were described in "Wild Geese and Eskimos" (1951), covering the Canadian trip and "A Thousand Geese" (1953) written jointly with James Fisher describing their experiences in Iceland. Both of these were illustrated by Peter Scott's drawings as well as many photographs.

Although neither are devoted exclusively to birds, Peter Scott's "Observations of Wildlife" (1980) and "Travel Diaries of a Naturalist" (1983) and (1985) are paragons of all that a naturalist's notebook should be, containing beautiful and accurate drawings accompanied by a readable and highly informative text.

Through his books, paintings and television programmes, Sir Peter Scott has given pleasure and instruction to countless numbers of people; whilst through his work for the Wildfowl Trust and World Wildlife Fund he has made a lasting and worldwide contribution to conservation and scientific ornithology.

Another eminent artist-naturalist, George Miksch Sutton was born in Bethany, Nebraska in May 1898. The son of a minister-teacher, the nature of his father's calling meant that the family had to move home frequently, and before he was eighteen George Sutton had lived in Oregon, Texas, Illinois and West Virginia. Whilst this nomadic existence may have interrupted his education somewhat, it certainly gave him the opportunity of seeing a great variety of birds, in which from an early age he showed a lively interest, particularly in sketching them. Whilst the family were living at Eugene, near the University of Oregon, Sutton was allowed to work in the University laboratory helping to arrange their birdskins in systematic order, but being only twelve years old at the time he was not permitted to use the collection!

In his later days he often affirmed that some of his happiest childhood memories were of when the family were living at Fort Worth in Texas; here he found, and reared, a nestling roadrunner, whose portrait, painted in 1913, is his earliest surviving drawing. The following year he raised two roadrunners, making a very thorough study of their behaviour and development. In January 1915 a painting of one of these appeared in "Bird Lore", the first of his published drawings.

At around this time he wrote to Louis Agassiz Fuertes, the great American bird painter. This was to be the beginning of a close friendship, and in 1916 George Sutton was invited to spend the summer with Louis Fuertes at Cayuga Lake near Ithaca, N.Y. Their correspondence extended over many years, and Fuertes' letters, full of advice and constructive criticism were later brought together by Sutton in a book entitled "To a

Young Bird Artist'' published in 1979.

During a visit to the Carnegie Museum in Pittsburgh George Sutton met Walter Edmund Clyde Todd, the Museum Curator of Birds, and in the summer of 1918 he was offered a temporary job at the museum helping with the egg collection and in repairing bird specimens. This was followed in 1919 by a full time job as Mr. Todd's assistant. An expert on the birds of Labrador, Clyde Todd had already made several expeditions to that Province, and in 1920 George Sutton accompanied him on yet another visit, which was to inspire much of his later work.

In 1923, by a special arrangement with Bethany College, he was allowed to take his degree long after he had studied there, and he later obtained a Ph.D. from Cornell University where, in 1932, he was appointed Curator of Birds. He continued to concentrate his studies on the birds of the far north, being the first man to discover the nest of Harris' sparrow, not far from Churchill on Hudsons Bay. He recorded his expedition to Churchill in his book ''Birds in the Wilderness, Adventures of an Ornithologist'' published in 1936 and which, like most of his books, he illustrated with his own paintings and drawings. A more technical account of which he was co-author with P.A. Tavener was ''The Birds of Churchill, Manitoba'' published in 1934 by the Carnegie Museum.

Over the years he travelled extensively throughout America, visiting Dakota, Saskatchewan and the panhandle of Oklahoma. In 1935 he took part in the joint Cornell University-American Museum of Natural History expedition to Florida and Louisiana where he had the opportunity, now almost certainly gone for ever, of sketching a pair of nesting ivory-billed woodpeckers. During the 1930s George Sutton became increasingly fascinated by Mexico and its birds, this led him to visit the country many times and resulted in three of his most attractive books: ''Mexican Birds — First Impressions'' (1951), ''At a bend in a Mexican River'' (1972) and ''Portraits of Mexican Birds'' (1975).

In the second world war George Sutton enrolled in the U.S.A.A.F., and because of his wide experience in the far north became chief of the Arctic section of "The Arctic, Desert and Tropic Information Center". Part of his time in the service was spent on Attu Island in the western Aleutians, an excellent place for observing birds. After leaving the service in 1947 he decided not to return to Cornell becoming instead Curator of Birds in the Museum of Zoology at the University of Michigan at Ann Arbor a post he held for some fourteen years. Oklahoma had been a favourite area of his and when in 1951 he was invited to work at the University of Oklahoma Biological Station at Lake Texoma he was delighted to accept. In 1952 he was asked to join the faculty as Professor of Zoology at the University of Oklahoma in Norman.

The far north never lost its enchantment for him and in 1958 he went to Iceland with Dr. Olin Sewell Pettingill. For him this was primarily a painting trip and his illustated account of this journey called "Arctic Summer: Adventures of a Bird Painter" was published in 1961. In 1962 he visited Victoria Island and in 1966 Jenny Lind Island, whilst on his last visit in 1969 he reached his furthest point north when he reached Cornwallis, Bathurst and Ellesmere Islands. His book "High Arctic" published in 1971 gives a fascinating account of these expeditions.

Besides his book he wrote numerous articles and more than 250 scientific papers. His close association with, and affection for, Oklahoma was celebrated in his long planned book "Oklahoma Birds" which was finally published in 1967.

A friendly and generous man, his death in December 1982 was keenly felt by his many friends, associates and students. His artistic talent was best described by Robert M. Mengel who wrote that he was "probably unique in being more ornithologist than any bird painter of equal note and more artistic than any other ornithologist of his stature". The importance of his work for ornithology in the U.S.A. was recognised by the bestowal of the Arthur A. Allen Award from Cornell University, the Distinguished Service Citation of the University of Oklahoma, and Knight Cross of the Order of The Falcon by the Icelandic Government.

Over the last fifty years the artist/author, Roger Tory Peterson, has probably made the greatest individual contribution to the practice of field identification. His "Field Guide to the Birds", originally published in 1934, marked the beginning of an entirely new concept in what could be termed "recognition" books. Earlier attempts to provide such a work had met with a limited success and, naturally, others have followed where he led, but overall the "Peterson system" remains paramount.

Roger Tory Peterson was born in Jamestown in New York State in August 1908, the son of poor Swedish parents, he left school at sixteen. From childhood he had been totally absorbed in the pursuit of natural history subjects, especially birds, spending most of his spare time in the countryside painting, studying and observing them. At the age of seventeen he attended the 1925 annual meeting of the American Ornithologists Union in New York and had two of his paintings accepted for the exhibition held during the conference. Even more importantly he met such men as Ludlow Griscom, Arthur Allen, Francis Lee Jaques and above all, Louis Fuertes, whose encouragement and advice fired his enthusiasm to become a professional ornithologist/artist.

He found a job, at eight dollars a week, painting chinese designs on the lacquerwork cabinets so popular at this time, until Willem Dieperink von Langereis, the head of the decorating department, highly impressed by one of Peterson's paintings, stongly advised him to go to an art school. This he did, enrolling, at Langereis' suggestion, at the Art Students League of New York in 1927. Speaking many years later he stated that "My primary contribution — field recognition — could not have been made had I followed the traditional path as a biologist. Because of my art background I approached things visually rather than phylogenetically, hence the Peterson field guide system was born".

In 1933 Peterson offered "A Field Guide to the Birds" to the publishers Houghton Mifflin who "decided to take a gamble" albeit a cautious one since they refused to offer any royalties on the first thousand copies and in the first instance, printed only two thousand. Mifflin had struck a goldmine, the first edition sold out almost at once, (a copy of this first printing, in mint condition with its jacket, would fetch more than a thousand dollars today). The book is currently in its fourth edition and more than two million copies have been sold to date. Over the years several improvements have been made, increasing the numbers of plates in colour — the present edition (1980) is illustrated entirely in colour — and the layout has been changed somewhat; originally the drawings and their relevant text had often been separated by several pages, but by and large the basic format has remained essentially unaltered.

A major reason for the immense success of this series is that by combining the roles of author and artist Peterson's work has a unity otherwise almost impossible to achieve. The "Peterson system", drawings of birds stressing their more important recognition features and using small arrows to draw attention to these particular points, show exactly what to look for when encountering an unfamiliar bird.

The three United States fieldguides covering the birds east of the 100th meridian, the Western birds and the birds of Texas are entirely Peterson's own work. In the case of "A Field Guide to the Birds of Britain and Europe" he had Guy Mountfort and Phillip Hollom as co-authors, and this, like its American counterpart, has gone into four editions. At the present time some twenty-five different fieldguides have appeared covering, besides birds, flowers, butterflies, mammals, reptiles, rocks and even the stars: whilst Peterson has not been actively involved in all these publications, all of them use his "system".

In 1934 Roger Tory Peterson took up an appointment with the National Audubon Society and eventually became its educational director and a senior lecturer. In latter years he has travelled and lectured widely as a birdleader for Lars Eric Lindblad on his ship the "Lindblad Explorer".

In addition to the fieldguide series Peterson is the author of "Birds over America" (1948), a very readable account of some of his earlier experiences and encounters in the course of numerous journeys around the United States in search of birds. His joint authorship with James Fisher of "Wild America" has already been mentioned, and a further work undertaken with him was "The World of Birds" published in 1964, which is an excellent treatise on birds and birdwatching, and beautifully illustrated by many of his pictures. A simplified edition of this book simply entitled "Birds" containing a new introduction by Peterson was published in 1971 after James Fisher's death.

Of the many honours and awards given to Roger Tory Peterson the most important is the Presidential Medal of Freedom, the highest civilian award of the U.S.A. He has, in addition, been granted nine honorary degrees from various universities, the Gold Medal of the World Wildlife Fund, the Gold Medal of the New York Zoological Society, the Arthur A. Allen Award and the William Brewster Award of the American Ornithologists Union.

Whilst this account has naturally concentrated on Peterson's ornithological achievements, some mention should be made of his botanical illustrations for his wild flower fieldguides as well as his beautiful paintings of birds with flower backgrounds. As artist, writer, lecturer and film maker through the range and impact of his activities Roger Tory Peterson has enlightened and encouraged literally millions of people.

The distributional atlas is a new idea in ornithological books which has developed steadily over the last forty years. Today it is generally acknowledged that a widely distributed force of observers, such as that deployed by the British Trust for Ornithology is essential for such enterprises, but early attempts by individuals at providing distribution maps go back at least to 1947 when James Fisher's "Bird Recognition" appeared. This excellent but uncompleted four volume work (of which three were published) pioneered the admirable "annual cycle diagrams" which show, at a glance, in what activities a species is engaged at any time of the year.

"Atlas of European Birds" by Professor Voous (1960) was a

worthwhile attempt to map distribution on a continental scale, and John Parslow in his book "Breeding Birds of Britain and Ireland" (1973) produced a comprehensive survey of the range and approximate populations of Britain's avifauna. These projects however which depended largely upon country and regional reports which showed merely that a species was present but giving no indication of its breeding status, suffered from a lack of detailed reports which can only be obtained from observers covering a small area on a regular basis. The first such ornithological atlas to be published in Britain was in 1970 when J. Lord and D.J. Munns in their "Atlas of Breeding Birds in the West Midlands" gave the breeding distribution of all species in seventy-three, ten kilometre squares, covering a census area in Worcestershire, Staffordshire and Warwickshire. This scale of ten kilometre squares had already been used, most successfully, by F.H. Perring and S.M. Walters in "The Atlas of the British Flora" published in 1962: and, after much discussion, the British Trust for Ornithology and the Irish Wildbird Conservancy in Eire decided to follow their example and adopt the same scale.

From 1968-1972 the greatest single survey so far attempted was carried out by between ten and fifteen thousand observers who submitted in total more than 285,000 record cards during the five breeding seasons covered by their field work. The ensuing book, "The Atlas of Breeding Birds in Britain and Ireland" edited by J.T.R. Sharrock was published in 1976. This survey only showed the presence and breeding status of species and did not indicate their numbers; however, in a follow-up survey of birds in winter at present under way it is intended that an estimate of the numbers of each species will be given.

"The Atlas of Breeding Birds in Britain and Ireland" was a forerunner of what will doubtless become a regular practice, not only on a national basis, but, as has already happened, on a county and regional basis. Other countries, Denmark, France, Holland and Switzerland amongst them, have followed suit, and in 1984 the Royal Australasian Ornithologists Union published "The Atlas of Australian Birds". Even a decade ago the idea of censusing so vast an area as Australia would have been inconceivable, but by dint of adopting areas of one degree of latitude by one of longitude, and some truly superhuman efforts on the part of some three thousand observers, the task was accomplished.

An interesting recent development in publishing is the return of books devoted to whole bird families. Popular at the turn of the century, for a time they virtually disappeared but are now making a comeback. The "new generation" monographs are of course not illustrated by hand coloured plates, but modern colour printing has produced some beautiful effects, and with a greatly improved text and taxonomy the resulting works are worthy successors to their Victorian and Edwardian predecessors.

One of the most lavish is "Kingfishers and Related Birds" by Joseph Forshaw, published in Australia in six volumes, two of which have been produced to date, a monograph on a truly heroic scale. "The Herons of the World" by James Hancock and Hugh Elliott (1978), "The Rails Of The World" by Dillon Ripley (1977), "The Birds of Paradise and Bower Birds" by Joseph Forshaw (1977) and "Eagles, Hawks and Falcons of the World" in two volumes (1968) by L. Brown and Dean Amadon are all recent examples and illustrated by fine coloured plates. Rather less comprehensively illustrated but nonetheless important works are "Pigeons and Doves of the World (1967), "Crows of the World" (1976) and "Estrildid Finches of the World" (1982) all by Derek Goodwin of the Sub-department of Ornithology of the British Museum (Nat. Hist.). In the United States Paul A. Johnsgard has written a number of important monographs such as "Ducks, Geese and Swans of the World" (1978), "The Plovers, Sandpipers and Snipes of the World" (1981), "The Grouse of the World" (1983) and "Cranes of the World" published in the same year. These are not illustrated in the "grand manner" but do utilise photographs and line drawings very effectively.

In addition to monographs of whole families, the publication of books devoted to a single species has continued and developed and today covers many birds of different families. "The Peregrine" by Derek Ratcliffe published in 1980 and "The Hen Harrier" by Donald Watson (1977) are excellent examples of this genre, among the seabirds "The Gannet" (1978) by Bryan Nelson and "The Puffin" (1984) by M.P. Harris have been the subject of detailed studies. That superb field observer Desmond Nethersole-Thompson has written four species monographs of birds which by reason of inaccessibility etc. are particularly difficult subjects for study. "The Greenshank" has featured in two of these books, one appearing in 1951 and a later one in 1979. "The Dotterel" was published in 1973 and "The Snow Bunting" in 1966, whilst his monograph on "Pine Crossbills" completes the quartet.

Since the war the increase in mobility and leisure together with the impact of natural history television programmes have impelled many people, who are not bird watchers in the accepted sense, to buy a bird book of some description. This interest has resulted in a flood of ornithological publications, some of very indifferent quality, especially many of those imported from Eastern Europe which are for the most part poorly produced and thoroughly unreliable. Whilst it is the fieldguides which have consistently achieved the biggest sales besides the Peterson Field Guide some other useful books have also been published. One of the earliest was "The Pocket Guide to British Birds" by Richard Fitter and illustrated by Richard Richardson, originally published in 1953 this still remains an excellent work and one which is especially helpful to beginners. "The Birds of Britain and Europe with North Africa and the Middle East" by Hermann Heinzel, Richard Fitter and John Parslow, published in 1972, covers a much wider area than the Peterson Field Guide and is of a consistently high quality as is its contemporary "The Hamlyn Guide to Birds of Britain and Europe" by Bertel Bruun and Arthur Singer.

Further guides to such previously remote areas as the Galapagos Islands, the Seychelles, Nepal and New Zealand have been produced to cater for the growing number of people able to take advantage of cheaper and easier air travel and the specialised birdwatching tours which are now widely available. Not only geographical fieldguides but guides to orders of birds have now appeared, and two books which come within this category are "Flight Identification of European Raptors" (1975) by R.F. Porter, I. Willis, S. Christensen and B.P. Nielson, and "Seabirds, An Identification Guide" (1983) by Peter Harrison.

Books on eggs are something of a rarity nowadays, but Dr. Colin Harrison of the British Museum (Nat. Hist.) has written "A Field Guide to the Nests, Eggs and Nestlings of British and European Birds" (1975) and "A Field Guide to the Nests, Eggs and Nestlings of Northern American Birds" (1978). In both these works the whole approach to the subject of birds' eggs is diametrically opposite to that adopted by earlier ornithologists, the emphasis being laid firmly on the identification of the eggs and not on their collection.

The manner in which ornithological knowledge has expanded in less than a century is well illustrated by the evolution of one book, Professor Alfred Newton's "A Dictionary of Birds" published originally in 1893-1896. When it was first printed this work encapsulated virtually everything that was then known about ornithology. To mark the centenary of the British Ornithologists Union — of which Professor Newton had been a founder, "A New Dictionary of Birds", edited by Sir Arthur Landsborough Thomson was published in 1964. In 1985 "A Dictionary of Birds" was again updated, this time under the editorship of Bruce Campbell and Elizabeth Lack. Professor Newton completed his dictionary with the assistance of Hans Gadow and three other contributors; Landsborough Thomson was able to call upon contributions from over two hundred authors and artists, while Bruce Campbell needed the services of more than 280 contributors, 18 artists and 44 photographers. In the second (1964) edition in addition to the very large number of new entries, virtually every entry, apart from some straight definitions of simple terms required amending: of the 1,230 entries in the original dictionary many consisted of definitions of local and ancient bird names besides numerous technical terms which had become obsolete. Although it was produced only twenty-one years later, the third edition needed an enormous amount of revision in order to take account of much subsequent research.

Much of the research being carried out today tends, at least in the first instance, to be reported in scientific journals, nevertheless a great many highly scientific works are added to the sum of ornithological literature every year. Amongst these "Avian Biology" edited by Donald Farner, J.R. King and Kenneth Parkes in the United States is an excellent example of current technical writing, each of the contributors being an acknowledged expert in his field. So far, between 1971 and 1983, they have filled seven weighty volumes, with more to come. The American counterpart of "A Dictionary of Birds" is "The Audubon Society Encyclopedia of North American Birds" by John K. Terres (1980). Using a very similar format it is written on a somewhat lower scientific level and its information is restricted to birds only found in the United States and Canada. One of the most beautifully produced books to have appeared in recent years, the quality of the photography and printing is quite exceptional. The text is concise and generally accurate and altogether this is an important work.

In the three hundred years which have elapsed between the publication of Willughby's first embryonic "handbook" and the "Handbook of the Birds of Europe, the Middle East and North Africa", attitudes have developed from uncritical wonderment laced with superstition, through exploration and investigation to a strict scientific discipline. There have been corresponding changes in literary styles, where the earlier more discursive type of writing has been superseded by the precise and often clinical approach adopted today: every work however is of some significance in that it adds to the sum total of ornithological knowledge. As for the authors themselves, from Giraldus Cambrensis to Sir Peter Scott, from John Ray to

Henry Seebohm, from William Yarrell to Roger Tory Peterson, they present a story of contrasts with characters of widely different natures, outlooks, temperaments and abilities, united only by an unwavering enthusiam for their subject and the contribution they have made to ornithological literature.

SELECT BIBLIOGRAPHY

Alexander, H.G. Seventy Years of Birdwatching. Poyser, 1974.

Allen, D.E. The Naturalist in Britain. Allen Lane, 1976.

Barber, L. The Heyday of Natural History. Jonathan Cape, 1980.

Breit V. & Clarke K.B. (comp.) Ornithology Books in the Library of Trinity College, Hartford. Trinity College, 1983.

Chapman, F.M. Autobiography of a Bird-Lover. Appleton-Century, 1933.

Cutright, P.R. & Brodhead, M.J. Elliot Coues, Naturalist and Frontier Historian. University of Illinois Press, 1981.

Devlin, J.C. & Naismith, G. The World of Roger Tory Peterson. New York Times Books, 1977.

Dictionary of National Biography. Oxford University Press London, 1975.

Farber, P.L. The Emergence of Ornithology as a Scientific Discipline, 1760-1850. Dordrecht, 1982.

Fisher, J. The Shell Bird Book. Ebury Press, 1966.

Fries, W.H. The Double Elephant Folio. The Story of Audubon's Birds of America. American Library Association. Chicago, 1973.

Guggisburg C.A.W. Early Wildlife Photographers. David & Charles, 1977.

Gurney, J.H. Early Annals of Ornithology. Witherby, 1921.

Jackson, C.E. Bird Illustrators. Witherby, 1975.

Jackson, C.E. Wood Engravings of Birds. Witherby, 1978.

McEvey, A. John Gould's Contribution to British Art. Sydney University Press, 1973.

Mullens W.H. & Kirke Swann H. A Bibliography of British Ornithology. Macmillan & Co., 1917.

Raven, C.E. John Ray, Naturalist C.U.P., 1942.

Raven, C.E. English Naturalists from Neckham to Ray. C.U.P., 1947.

Rothschild, M. Dear Lord Rothschild. Hutchinson, 1983.

Sauer, G.C. John Gould, The Birdman. Sotheran, 1982.
Stresemann, E. Ornithology From Aristotle To the Present Day. Harvard University Press, 1975.
Sutton, G.M. Bird Student. University of Texas Press, 1980.
Thomas, K. Man and the Natural World. Allen Lane, 1983.
Welker, R.H. Birds and Men. Belknap Press of Harvard University Press, Cambridge, 1955.
Wood, Casey A. An Introduction to the Literature of Vertebrate Zoology. Oxford, 1931.
Zimmer, J.T. Catalogue of the Edward E. Ayer Library. Chicago, 1926.

JOURNALS

The Auk.
American Naturalist.
"British Birds".
The Ibis.
The Journal of the Society for the Bibliography of Natural History (Later Archives of Natural History).
The Living Bird.

Index

All book titles are listed under the name of the author or editor, and in the case of multiple authors under the first named on the title-page. Books listed in the index do not necessarily form a complete bibliography of an author's work; and in certain cases long subtitles have not been repeated.

Index

Index